Lecture Notes in Physics

New Series m: Monographs

Springer
Berlin
Heidelberg
New York
Barcelona
Budapest
Hong Kong
London
Milan
Paris
Santa Clara
Singapore
Tokyo

The Editorial Policy for Monographs

The series Lecture Notes in Physics reports new developments in physical research and teaching - quickly, informally, and at a high level. The type of material considered for publication in the New Series m includes monographs presenting original research or new angles in a classical field. The timeliness of a manuscript is more important than its form, which may be preliminary or tentative. Manuscripts should be reasonably self-contained. They will often present not only results of the author(s) but also related work by other people and will provide sufficient motivation, examples, and applications.
The manuscripts or a detailed description thereof should be submitted either to one of the series editors or to the managing editor. The proposal is then carefully refereed. A final decision concerning publication can often only be made on the basis of the complete manuscript, but otherwise the editors will try to make a preliminary decision as definite as they can on the basis of the available information.
Manuscripts should be no less than 100 and preferably no more than 400 pages in length. Final manuscripts should preferably be in English, or possibly in French or German. They should include a table of contents and an informative introduction accessible also to readers not particularly familiar with the topic treated. Authors are free to use the material in other publications. However, if extensive use is made elsewhere, the publisher should be informed. Authors receive jointly 50 complimentary copies of their book. They are entitled to purchase further copies of their book at a reduced rate. As a rule no reprints of individual contributions can be supplied. No royalty is paid on Lecture Notes in Physics volumes. Commitment to publish is made by letter of interest rather than by signing a formal contract. Springer-Verlag secures the copyright for each volume.

The Production Process

The books are hardbound, and quality paper appropriate to the needs of the author(s) is used. Publication time is about ten weeks. More than twenty years of experience guarantee authors the best possible service. To reach the goal of rapid publication at a low price the technique of photographic reproduction from a camera-ready manuscript was chosen. This process shifts the main responsibility for the technical quality considerably from the publisher to the author. We therefore urge all authors to observe very carefully our guidelines for the preparation of camera-ready manuscripts, which we will supply on request. This applies especially to the quality of figures and halftones submitted for publication. Figures should be submitted as originals or glossy prints, as very often Xerox copies are not suitable for reproduction. For the same reason, any writing within figures should not be smaller than 2.5 mm. It might be useful to look at some of the volumes already published or, especially if some atypical text is planned, to write to the Physics Editorial Department of Springer-Verlag direct. This avoids mistakes and time-consuming correspondence during the production period.
As a special service, we offer free of charge LaTeX and TeX macro packages to format the text according to Springer-Verlag's quality requirements. We strongly recommend authors to make use of this offer, as the result will be a book of considerably improved technical quality.
Manuscripts not meeting the technical standard of the series will have to be returned for improvement.
For further information please contact Springer-Verlag, Physics Editorial Department II, Tiergartenstrasse 17, D-69121 Heidelberg, Germany.

Giovanni Landi

An Introduction
to Noncommutative Spaces
and Their Geometries

 Springer

Author

Giovanni Landi
Dipartimento di Scienze Matematiche
Università degli Studi di Trieste
P. le Europa, 1
I-34127 Trieste, Italy

CIP data applied for

Die Deutsche Bibliothek - CIP-Einheitsaufnahme

Landi, Giovanni:
An introduction to noncommutative spaces and their geometries /
Giovanni Landi. - Berlin ; Heidelberg ; New York ; Barcelona ;
Budapest ; Hong Kong ; London ; Milan ; Paris ; Santa Clara ;
Singapore ; Tokyo : Springer, 1997
  (Lecture notes in physics : N.s. M, Monographs ; 51)
  ISBN 3-540-63509-2

ISSN 0940-7677 (Lecture Notes in Physics. New Series m: Monographs)
ISBN 3-540-63509-2 Edition Springer-Verlag Berlin Heidelberg New York

Typesetting: Camera-ready by author
Cover design: *design & production* GmbH, Heidelberg
SPIN: 10550641   55/3144-543210 - Printed on acid-free paper

ad Anna e Jacopo
per il loro amore e la loro pazienza

Preface

These notes arose from a series of introductory seminars on noncommutative geometry I gave at the University of Trieste in September 1995 during the X Workshop on Differential Geometric Methods in Classical Mechanics. It was Beppe Marmo's suggestion that I wrote notes for the lectures.

The notes are mainly an introduction to Connes' noncommutative geometry. They could serve as a 'first aid kit' before one ventures into the beautiful but bewildering landscape of Connes' theory. The main difference from other available introductions to Connes' work, notably Kastler's papers [86] and also the Gracia-Bondía and Varilly paper [130], is the emphasis on noncommutative spaces seen as concrete spaces.

Important examples of noncommutative spaces are provided by noncommutative lattices. The latter are the subject of intense work I am doing in collaboration with A.P. Balachandran, Giuseppe Bimonte, Elisa Ercolessi, Fedele Lizzi, Gianni Sparano and Paulo Teotonio-Sobrinho. These notes are also meant to be an introduction to this research. There is still a lot of work in progress and by no means can these notes be considered as a review of everything we have achieved so far. Rather, I hope they will show the relevance and potentiality for physical theories of noncommutative lattices.

Acknowledgement.
I am indebted to several people for help and suggestions of different kinds at various stages of this project: A.P. Balachandran, G. Bimonte, U. Bruzzo, T. Brzezinski, M. Carfora, R. Catenacci, A. Connes, L. Dabrowski, G.F. Dell'Antonio, M. Dubois-Violette, B. Dubrovin, E. Elizalde, E. Ercolessi, J.M. Gracia-Bondía, P. Hajac, D. Kastler, A. Kempf, F. Lizzi, J. Madore, G. Marmo, A. Napoli, C. Reina, C. Rovelli, G. Sewell, P. Siniscalco, G. Sparano, P. Teotonio-Sobrinho, G. Thompson, J.C. Várilly, R. Zapatrin.

Contents

1. Introduction

In the last fifteen years, there has been an increasing interest in noncommutative (and/or quantum) geometry both in mathematics and in physics.

In A. Connes' functional analytic approach [32], noncommutative C^*-algebras are the 'dual' arena for noncommutative topology. The (commutative) Gel'fand-Naimark theorem (see for instance [65]) states that there is a complete equivalence between the category of (locally) compact Hausdorff spaces and (proper and) continuous maps and the category of commutative (not necessarily) unital[1] C^*-algebras and *-homomorphisms. Any commutative C^*-algebra can be realized as the C^*-algebra of complex valued functions over a (locally) compact Hausdorff space. A noncommutative C^*-algebra will now be thought of as the algebra of continuous functions on some 'virtual noncommutative space'. The attention will be switched from spaces, which in general do not even exist 'concretely', to algebras of functions.

Connes has also developed a new calculus, which replaces the usual differential calculus. It is based on the notion of a real spectral triple $(\mathcal{A}, \mathcal{H}, D, J)$ where $\mathcal{A}$ is a noncommutative *-algebra (indeed, in general not necessarily a C^*-algebra), $\mathcal{H}$ is a Hilbert space on which $\mathcal{A}$ is realized as an algebra of bounded operators, and D is an operator on $\mathcal{H}$ with suitable properties and which contains (almost all) the 'geometric' information. The antilinear isometry J on $\mathcal{H}$ will provide a real structure on the triple. With any closed n-dimensional Riemannian spin manifold M there is associated a canonical spectral triple with $\mathcal{A} = C^\infty(M)$, the algebra of complex valued smooth functions on M; $\mathcal{H} = L^2(M, S)$, the Hilbert space of square integrable sections of the irreducible spinor bundle over M; and D the Dirac operator associated with the Levi-Civita connection. For this triple Connes' construction gives back the usual differential calculus on M. In this case J is the composition of the charge conjugation operator with usual complex conjugation.

Yang-Mills and gravity theories stem from the notion of connection (gauge or linear) on vector bundles. The possibility of extending these notions to the realm of noncommutative geometry relies on another classical duality. The Serre-Swan theorem [123] states that there is a complete equivalence between the category of (smooth) vector bundles over a (smooth) compact space and bundle maps and the category of projective modules of finite type over com-

---

[1] A unital C^*-algebras is a C^*-algebras which has a unit, see Sect. 2.1.

mutative algebras and module morphisms. The space $\Gamma(E)$ of (smooth) sections of a vector bundle E over a compact space is a projective module of finite type over the algebra $C(M)$ of (smooth) functions over M and any finite projective $C(M)$-module can be realized as the module of sections of some bundle over M.

With a noncommutative algebra $\mathcal{A}$ as the starting ingredient, the (analogue of) vector bundles will be projective modules of finite type over $\mathcal{A}$.[2] One then develops a full theory of connections which culminates in the definition of a Yang-Mills action. Needless to say, starting with the canonical triple associated with an ordinary manifold one recovers the usual gauge theory. But now, one has a much more general setting. In [38] Connes and Lott computed the Yang-Mills action for a space $M \times Y$ which is the product of a Riemannian spin manifold M by a 'discrete' internal space Y consisting of two points. The result is a Lagrangian which reproduces the Standard Model with its Higgs sector with quartic symmetry breaking self-interaction and the parity violating Yukawa coupling with fermions. A nice feature of the model is a geometric interpretation of the Higgs field which appears as the component of the gauge field in the internal direction. Geometrically, the space $M \times Y$ consists of two sheets which are at a distance of the order of the inverse of the mass scale of the theory. Differentiation on $M \times Y$ consists of differentiation on each copy of M together with a finite difference operation in the Y direction. A gauge potential A decomposes as a sum of an ordinary differential part $A^{(1,0)}$ and a finite difference part $A^{(0,1)}$ which gives the Higgs field.

Quite recently Connes [36] has proposed a pure 'geometrical' action which, for a suitable noncommutative algebra $\mathcal{A}$ (noncommutative geometry of the Standard Model), yields the Standard Model Lagrangian coupled with Einstein gravity. The group $Aut(\mathcal{A})$ of automorphisms of the algebra plays the rôle of the diffeomorphism group while the normal subgroup $Inn(\mathcal{A}) \subseteq Aut(\mathcal{A})$ of inner automorphisms gives the gauge transformations. Internal fluctuations of the geometry, produced by the action of inner automorphisms, give the gauge degrees of freedom.

A theory of linear connections and Riemannian geometry, culminating in the analogue of the Hilbert-Einstein action in the context of noncommutative geometry has been proposed in [26]. Again, for the canonical triple one recovers the usual Einstein gravity. When computed for a Connes-Lott space $M \times Y$ as in [26], the action produces a Kaluza-Klein model which contains the usual integral of the scalar curvature of the metric on M, a minimal coupling for the scalar field to such a metric, and a kinetic term for the scalar field. A somewhat different model of geometry on the space $M \times Y$ produces an action which is just the Kaluza-Klein action of unified

---

[2] In fact, the generalization is not so straightforward, see Chapter 4 for a better discussion.

gravity-electromagnetism consisting of the usual gravity term, a kinetic term for a minimally coupled scalar field and an electromagnetic term [95].

Algebraic K-theory of an algebra $\mathcal{A}$, as the study of equivalence classes of projective modules of finite type over $\mathcal{A}$, provides analogues of topological invariants of the 'corresponding virtual spaces'. On the other hand, cyclic cohomology provides analogues of differential geometric invariants. K-theory and cohomology are connected by the Chern character. This has found a beautiful application by Bellissard [9] to the quantum Hall effect. He has constructed a natural cyclic 2-cocycle on the noncommutative algebra of function on the Brillouin zone. The Hall conductivity is just the pairing between this cyclic 2-cocycle and an idempotent in the algebra: the spectral projection of the Hamiltonian. A crucial rôle in this analysis is played by the noncommutative torus [115].

In these notes we give a self-contained introduction to a limited part of Connes' noncommutative theory, without even trying to cover all its aspects. Our main objective is to present some of the physical applications of noncommutative geometry.

In Chapter 2, we introduce C^*-algebras and the (commutative) Gel'fand-Naimark theorem. We then move to structure spaces of noncommutative C^*-algebras. We describe to some extent the space $Prim\mathcal{A}$ of an algebra $\mathcal{A}$ with its natural Jacobson topology. Examples of such spaces turn out to be relevant in an approximation scheme to 'continuum' topological spaces by means of projective systems of lattices with a nontrivial T_0 topology [121]. Such lattices are truly noncommutative lattices since their algebras of continuous functions are noncommutative C^*-algebras of operator valued functions. Techniques from noncommutative geometry have been used to construct models of gauge theory on these noncommutative lattices [6, 7]. Noncommutative lattices are described at length in Chapter 3.

In Chapter 4 we describe the theory of projective modules and the Serre-Swan theorem. Then we develop the notion of Hermitian structure, an algebraic counterpart of a metric. We also mention other relevant categories of (bi)modules such as central and diagonal bimodules. Following this, in Section 5 we provide a few fundamentals of K-theory. As an example, we describe at length the K-theory of the algebra of the Penrose tiling of the plane.

Chapter 6 is devoted to the theory of infinitesimals and the spectral calculus. We first describe the Dixmier trace which plays a fundamental rôle in the theory of integration. Then the notion of a spectral triple is introduced with the associated definition of distance and integral on a 'noncommutative space'. We work out in detail the example of the canonical triple associated with any Riemannian spin manifold. Noncommutative forms are then introduced in Chapter 7. Again, we show in detail how to recover the usual exterior calculus of forms.

In the first part of Chapter 8, we describe abelian gauge theories in order to get some feeling for the structures. We then develop the theory of connections, compatible connections, and gauge transformations.

In Chapters 9 and 10 we describe field theories on modules. In particular, in Chapter 9 we show how to construct Yang-Mills and fermionic models. Gravity models are treated in Chapter 10. In Chapter 11 we describe a simple quantum mechanical system on a noncommutative lattice, namely the θ-quantization of a particle on a noncommutative lattice approximating the circle.

We feel we should warn the interested reader that we shall not give any detailed account of the construction of the standard model in noncommutative geometry nor of the use of the latter for model building in particle physics. We shall limit ourselves to a very sketchy overview while referring to the existing and rather useful literature on the subject.

The appendices contain material related to the ideas developed in the text.

As alluded to before, the territory of noncommutative and quantum geometry is so vast and new regions are discovered at such a high speed that the number of relevant papers is overwhelming. It is impossible to even think of covering 'everything'. We just finish this introduction with a partial list of references for 'further reading'. The generalization from classical (differential) geometry to noncommutative (differential) geometry is not unique. This is a consequence of the existence of several types of noncommutative algebras. A direct noncommutative generalization of the algebraic approach of Koszul [89] to differential geometry is given by the so-called 'derivation-based calculus' proposed in [47]. Given a noncommutative algebra $\mathcal{A}$ one takes as the analogue of vector fields the Lie algebra $Der\mathcal{A}$ of derivations of $\mathcal{A}$. Besides the fact that, due to noncommutativity, $Der\mathcal{A}$ is a module only over the center of $\mathcal{A}$, there are several algebras which admit only few derivations. However, if we think of $\mathcal{A}$ as replacing the algebra of smooth functions on a manifold, the derivation based calculus is 'natural' in the sense that it depends only on $\mathcal{A}$ and does not require additional structures (although, in a sense, one is fixing 'a priori' a smooth structure). We refer to [50, 100] for the details and several applications to Yang-Mills models and gravity theories. Here we only mention that this approach fits well with quantum mechanics [48, 49]: since derivations are infinitesimal algebra automorphisms, they are natural candidates for differential evolution equations and noncommutative dynamical systems, notably classical and quantum mechanical systems. In [29, 39, 116] a calculus, with derivations related to a group action in the framework of C^*-dynamical systems, has been used to construct a noncommutative Yang-Mills theory on noncommutative tori [115]. In [91, 92] (see also references therein) a calculus, with derivations for commutative algebras, together with extensions of Lie algebras of derivations, has been used to construct algebraic gauge theories. Furthermore, algebraic gravity models have been constructed by generalizing the notion of Einstein algebras [71].

In [133] noncommutative geometry was used to formulate the classical field theory of strings (see also [75]). For Hopf algebras and quantum groups and their applications to quantum field theory we refer to [46, 69, 78, 84, 99, 111, 124]. Twisted (or pseudo) groups have been proposed in [135]. For other interesting quantum spaces such as the quantum plane we refer to [101] and [132]. Very interesting work on the structure of space-time has been done in [45, 51]. We also mention the work on infrared and ultraviolet regularizations in [88].

The reference for Connes' noncommutative geometry is 'par excellence' his book [32]. The paper [130] has also been very helpful.

2. Noncommutative Spaces
and Algebras of Functions

The starting idea of noncommutative geometry is the shift from spaces to algebras of functions defined on them. In general, one has only the algebra and there is no analogue of space whatsoever. In this Chapter we shall give some general facts about algebras of (continuous) functions on (topological) spaces. In particular we shall try to make some sense of the notion of a 'noncommutative space'.

2.1 Algebras

Here we present mainly the objects that we shall need later on while referring to [19, 43, 110] for details. In the sequel, any algebra $\mathcal{A}$ will be an algebra over the field of complex numbers $\mathbb{C}$. This means that $\mathcal{A}$ is a vector space over $\mathbb{C}$ so that objects like $\alpha a + \beta b$ with $a, b \in \mathcal{A}$ and $\alpha, \beta \in \mathbb{C}$, make sense. Also, there is a product $\mathcal{A} \times \mathcal{A} \to \mathcal{A}$, $\mathcal{A} \times \mathcal{A} \ni (a, b) \mapsto ab \in \mathcal{A}$, which is distributive over addition,

$$a(b + c) = ab + ac , \quad (a + b)c = ac + bc , \quad \forall\, a, b, c \in \mathcal{A} . \tag{2.1}$$

In general, the product is not commutative so that

$$ab \neq ba . \tag{2.2}$$

We shall assume that $\mathcal{A}$ has a unit, namely an element $\mathbb{I}$ such that

$$a\mathbb{I} = \mathbb{I}a , \quad \forall\, a \in \mathcal{A} . \tag{2.3}$$

On occasion we shall comment on the situations for which this is not the case. An algebra with a unit will also be called a *unital* algebra.

The algebra $\mathcal{A}$ is called a **-algebra* if it admits an (antilinear) involution $* : \mathcal{A} \to \mathcal{A}$ with the properties,

$$
\begin{aligned}
a^{**} &= a , \\
(ab)^* &= b^* a^* , \\
(\alpha a + \beta b)^* &= \overline{\alpha} a^* + \overline{\beta} b^* ,
\end{aligned}
\tag{2.4}
$$

for any $a, b \in \mathcal{A}$ and $\alpha, \beta \in \mathbb{C}$ and bar denotes usual complex conjugation. A *normed algebra* $\mathcal{A}$ is an algebra with a norm $||\cdot|| : \mathcal{A} \to \mathbb{R}$ which has the properties,

$$\begin{aligned}
&||a|| \geq 0 , \quad ||a|| = 0 \Leftrightarrow a = 0 , \\
&||\alpha a|| = |\alpha| \, ||a||, \\
&||a + b|| \leq ||a|| + ||b||, \\
&||ab|| \leq ||a|| \, ||b||,
\end{aligned} \tag{2.5}$$

for any $a, b \in \mathcal{A}$ and $\alpha \in \mathbb{C}$. The third condition is called the triangle inequality while the last one is called the product inequality. The topology defined by the norm is called the *norm* or *uniform topology*. The corresponding neighborhoods of any $a \in \mathcal{A}$ are given by

$$U(a, \varepsilon) = \{b \in \mathcal{A} \mid ||a - b|| < \varepsilon\} , \quad \varepsilon > 0 . \tag{2.6}$$

A *Banach algebra* is a normed algebra which is complete in the uniform topology.

A *Banach* $*$-*algebra* is a normed $*$-algebra which is complete and such that

$$||a^*|| = ||a||, \quad \forall \, a \in \mathcal{A} . \tag{2.7}$$

A C^*-*algebra* $\mathcal{A}$ is a Banach $*$-algebra whose norm satisfies the additional identity

$$||a^*a|| = ||a||^2, \quad \forall \, a \in \mathcal{A} . \tag{2.8}$$

In fact, this property, together with the product inequality yields (2.7) automatically. Indeed, $||a||^2 = ||a^*a|| \leq ||a^*|| \, ||a||$ from which $||a|| \leq ||a^*||$. By interchanging a with a^* one gets $||a^*|| \leq ||a||$ and in turn (2.7).

Example 2.1.1. The commutative algebra $\mathcal{C}(M)$ of continuous functions on a compact Hausdorff topological space M, with $*$ denoting complex conjugation and the norm given by the supremum norm,

$$||f||_\infty = \sup_{x \in M} |f(x)| , \tag{2.9}$$

is an example of commutative C^*-algebra. If M is not compact but only locally compact, then one should take the algebra $\mathcal{C}_0(M)$ of continuous functions vanishing at infinity; this algebra has no unit. Clearly $\mathcal{C}(M) = \mathcal{C}_0(M)$ if M is compact. One can prove that $\mathcal{C}_0(M)$ (and a fortiori $\mathcal{C}(M)$ if M is compact) is complete in the supremum norm.[1]

---

[1] We recall that a function $f : M \to \mathbb{C}$ on a locally compact Hausdorff space is said to *vanish at infinity* if for every $\epsilon > 0$ there exists a compact set $K \subset M$ such that $|f(x)| < \epsilon$ for all $x \notin K$. As mentioned in App. A.1, the algebra $\mathcal{C}_0(M)$ is the closure in the norm (2.9) of the algebra of functions with compact support. The function f is said to have compact support if the space $K_f =: \{x \in M \mid f(x) \neq 0\}$ is compact [118].

Example 2.1.2. The noncommutative algebra $\mathcal{B}(\mathcal{H})$ of bounded linear operators on an infinite dimensional Hilbert space $\mathcal{H}$ with involution * given by the adjoint and the norm given by the operator norm,

$$||B|| = \sup\{||B\chi|| : \chi \in \mathcal{H}, ||\chi|| \leq 1\} , \qquad (2.10)$$

gives a noncommutative C^*-algebra.

Example 2.1.3. As a particular case of the previous, consider the noncommutative algebra $\mathbb{M}_n(\mathbb{C})$ of $n \times n$ matrices T with complex entries, with T^* given by the Hermitian conjugate of T. The norm (2.10) can also be equivalently written as

$$||T|| = \text{ the positive square root of the largest eigenvalue of } T^*T . \quad (2.11)$$

On the algebra $\mathbb{M}_n(\mathbb{C})$ one could also define a different norm,

$$||T||' = sup\{T_{ij}\} , \quad T = (T_{ij}) . \qquad (2.12)$$

One can easily realize that this norm is not a C^*-norm, the property (2.8) not being fulfilled. It is worth noticing though, that the two norms (2.11) and (2.12) are equivalent as Banach norms in the sense that they define the same topology on $\mathbb{M}_n(\mathbb{C})$: any ball in the topology of the norm (2.11) is contained in a ball in the topology of the norm (2.12) and viceversa.

A (proper, norm closed) subspace $\mathcal{I}$ of the algebra $\mathcal{A}$ is a *left ideal* (respectively a *right ideal*) if $a \in \mathcal{A}$ and $b \in \mathcal{I}$ imply that $ab \in \mathcal{I}$ (respectively $ba \in \mathcal{I}$). A *two-sided ideal* is a subspace which is both a left and a right ideal. The ideal $\mathcal{I}$ (left, right or two-sided) is called *maximal* if there exists no other ideal of the same kind in which $\mathcal{I}$ is contained. Each ideal is automatically an algebra. If the algebra $\mathcal{A}$ has an involution, any *-ideal (namely an ideal which contains the * of any of its elements) is automatically two-sided. If $\mathcal{A}$ is a Banach *-algebra and $\mathcal{I}$ is a two-sided *-ideal which is also closed (in the norm topology), then the quotient $\mathcal{A}/\mathcal{I}$ can be made into a Banach *-algebra. Furthermore, if $\mathcal{A}$ is a C^*-algebra, then the quotient $\mathcal{A}/\mathcal{I}$ is also a C^*-algebra. The C^*-algebra $\mathcal{A}$ is called *simple* if it has no nontrivial two-sided ideals. A two-sided ideal $\mathcal{I}$ in the C^*-algebra $\mathcal{A}$ is called *essential* in $\mathcal{A}$ if any other non-zero ideal in $\mathcal{A}$ has a non-zero intersection with it.

If $\mathcal{A}$ is any algebra, the *resolvent set* $r(a)$ of an element $a \in \mathcal{A}$ is the subset of complex numbers given by

$$r(a) = \{\lambda \in \mathbb{C} \mid a - \lambda \mathbb{I} \text{ is invertible}\} . \qquad (2.13)$$

For any $\lambda \in r(a)$, the inverse $(a - \lambda \mathbb{I})^{-1}$ is called the *resolvent* of a at λ. The complement of $r(a)$ in $\mathbb{C}$ is called the *spectrum* $\sigma(a)$ of a. While for a general algebra, the spectra of its elements may be rather complicated, for C^*-algebras they are quite nice. If $\mathcal{A}$ is a C^*-algebra, it turns out that the spectrum of any of its element a is a nonempty compact subset of $\mathbb{C}$. The *spectral radius* $\rho(a)$ of $a \in \mathcal{A}$ is given by

$$\rho(a) = sup\{|\lambda| \ , \ \lambda \in \sigma(a)\} \tag{2.14}$$

and, $\mathcal{A}$ being a C^*-algebra, it turns out that

$$||a||^2 = \rho(a^*a) =: sup\{|\lambda| \ | \ a^*a - \lambda \ \text{not invertible} \ \} \ , \quad \forall \, a \in \mathcal{A} \ . \tag{2.15}$$

A C^*-algebra is really such for a unique norm given by the spectral radius as in (2.15): the norm is uniquely determined by the algebraic structure.

An element $a \in \mathcal{A}$ is called *self-adjoint* if $a = a^*$. The spectrum of any such element is real and $\sigma(a) \subseteq [\ -||a||, ||a|| \ ]$, $\sigma(a^2) \subseteq [0, ||a||^2]$. An element $a \in \mathcal{A}$ is called *positive* if it is self-adjoint and its spectrum is a subset of the positive half-line. It turns out that the element a is positive if and only if $a = b^*b$ for some $b \in \mathcal{A}$. If $a \neq 0$ is positive, one also writes $a > 0$.

A **-morphism* between two C^*-algebras $\mathcal{A}$ and $\mathcal{B}$ is any $\mathbb{C}$-linear map $\pi : \mathcal{A} \to \mathcal{B}$ which in addition is a *homomorphism* of algebras, namely, it satisfies the multiplicative condition,

$$\pi(ab) = \pi(a)\pi(b) \ , \quad \forall \, a, b \in \mathcal{A} \ . \tag{2.16}$$

and is *-preserving,

$$\pi(a^*) = \pi(a)^* \ , \quad \forall \, a \in \mathcal{A} \ . \tag{2.17}$$

These conditions automatically imply that π is positive, $\pi(a) \geq 0$ if $a \geq 0$. Indeed, if $a \geq 0$, then $a = b^*b$ for some $b \in \mathcal{A}$; as a consequence, $\pi(a) = \pi(b^*b) = \pi(b)^*\pi(b) \geq 0$. It also turns out that π is automatically continuous, norm decreasing,

$$||\pi(a)||_{\mathcal{B}} \leq ||a||_{\mathcal{A}} \ , \quad \forall \, a \in \mathcal{A} \ , \tag{2.18}$$

and the image $\pi(\mathcal{A})$ is a C^*-subalgebra of $\mathcal{B}$. A *-morphism π which is also bijective as a map, is called a **-isomorphism* (the inverse map π^{-1} is automatically a *-morphism).

A *representation* of a C^*-algebra $\mathcal{A}$ is a pair $(\mathcal{H}, \pi)$ where $\mathcal{H}$ is a Hilbert space and π is a *-morphism

$$\pi : \mathcal{A} \longrightarrow \mathcal{B}(\mathcal{H}) \ , \tag{2.19}$$

with $\mathcal{B}(\mathcal{H})$ the C^*-algebra of bounded operators on $\mathcal{H}$.

The representation $(\mathcal{H}, \pi)$ is called *faithful* if $ker(\pi) = \{0\}$, so that π is a *-isomorphism between $\mathcal{A}$ and $\pi(\mathcal{A})$. One can prove that a representation is faithful if and only if $||\pi(a)|| = ||a||$ for any $a \in \mathcal{A}$ or $\pi(a) > 0$ for all $a > 0$. The representation $(\mathcal{H}, \pi)$ is called *irreducible* if the only closed subspaces of $\mathcal{H}$ which are invariant under the action of $\pi(\mathcal{A})$ are the trivial subspaces $\{0\}$ and $\mathcal{H}$. One proves that a representation is irreducible if and only if the commutant $\pi(\mathcal{A})'$ of $\pi(\mathcal{A})$, i.e. the set of of elements in $\mathcal{B}(\mathcal{H})$ which commute with each element in $\pi(\mathcal{A})$, consists of multiples of the identity operator. Two representations $(\mathcal{H}_1, \pi_1)$ and $(\mathcal{H}_2, \pi_2)$ are said to be *equivalent* (or more precisely, *unitary equivalent*) if there exists a unitary operator $U : \mathcal{H}_1 \to \mathcal{H}_2$, such that

$$\pi_1(a) = U^* \pi_2(a) U , \quad \forall \, a \in \mathcal{A} . \tag{2.20}$$

In App. A.2 we describe the notion of states of a C^*-algebra and the representations associated with them via the Gel'fand-Naimark-Segal construction.

A subspace $\mathcal{I}$ of the C^*-algebra $\mathcal{A}$ is called a *primitive ideal* if it is the kernel of an irreducible representation, namely $\mathcal{I} = ker(\pi)$ for some irreducible representation $(\mathcal{H}, \pi)$ of $\mathcal{A}$. Notice that $\mathcal{I}$ is automatically a two-sided ideal which is also closed. If $\mathcal{A}$ has a faithful irreducible representation on some Hilbert space so that the set $\{0\}$ is a primitive ideal, it is called a *primitive C^*-algebra*. The set $Prim\mathcal{A}$ of all primitive ideals of the C^*-algebra $\mathcal{A}$ will play a crucial rôle in following Chapters.

2.2 Commutative Spaces

The content of the commutative Gel'fand-Naimark theorem is precisely the fact that given *any* commutative C^*-algebra $\mathcal{C}$, one can reconstruct a Hausdorff[2] topological space M such that $\mathcal{C}$ is isometrically $*$-isomorphic to the algebra of (complex valued) continuous functions $\mathcal{C}(M)$ [43, 65].

In this Section $\mathcal{C}$ denotes a fixed commutative C^*-algebra with unit. Given such a $\mathcal{C}$, we let $\widehat{\mathcal{C}}$ denote the *structure space* of $\mathcal{C}$, namely the space of equivalence classes of irreducible representations of $\mathcal{C}$. The trivial representation, given by $\mathcal{C} \to \{0\}$, is not included in $\widehat{\mathcal{C}}$. The C^*-algebra $\mathcal{C}$ being commutative, every irreducible representation is one-dimensional. It is then a (non-zero) $*$-linear functional $\phi : \mathcal{C} \to \mathbb{C}$ which is multiplicative, i.e. it satisfies $\phi(ab) = \phi(a)\phi(b)$ for any $a, b \in \mathcal{C}$. It follows that $\phi(\mathbb{I}) = 1$, $\forall \, \phi \in \widehat{\mathcal{C}}$. Any such multiplicative functional is also called a *character* of $\mathcal{C}$. Then, the space $\widehat{\mathcal{C}}$ is also the space of all characters of $\mathcal{C}$.

The space $\widehat{\mathcal{C}}$ is made into a topological space, called the *Gel'fand space* of $\mathcal{C}$, by endowing it with the *Gel'fand topology*, namely the topology of pointwise convergence on $\mathcal{C}$. A sequence $\{\phi_\lambda\}_{\lambda \in \Lambda}$ (Λ is any directed set) of elements of $\widehat{\mathcal{C}}$ converges to $\phi \in \widehat{\mathcal{C}}$ if and only if for any $c \in \mathcal{C}$, the sequence $\{\phi_\lambda(c)\}_{\lambda \in \Lambda}$ converges to $\phi(c)$ in the topology of $\mathbb{C}$. The algebra $\mathcal{C}$ having a unit implies $\widehat{\mathcal{C}}$ is a compact Hausdorff space. The space $\widehat{\mathcal{C}}$ is only locally compact if $\mathcal{C}$ is without a unit.

---

[2] We recall that a topological space is called Hausdorff if for any two points of the space there are two open disjoint neighborhoods each containing one of the points [59].

Equivalently, $\widehat{C}$ could be taken to be the space of maximal ideals (automatically two-sided) of C instead of the space of irreducible representations.[3] Since the C^*-algebra C is commutative these two constructions agree because, on one side, kernels of (one-dimensional) irreducible representations are maximal ideals, and, on the other side, any maximal ideal is the kernel of an irreducible representation [65]. Indeed, consider $\phi \in \widehat{C}$. Then, since $C = Ker(\phi) \oplus \mathbb{C}$, the ideal $Ker(\phi)$ is of codimension one and so it is a maximal ideal of C. Conversely, suppose that $\mathcal{I}$ is a maximal ideal of C. Then, the natural representation of C on $C/\mathcal{I}$ is irreducible, hence one-dimensional. It follows that $C/\mathcal{I} \cong \mathbb{C}$, so that the quotient homomorphism $C \to C/\mathcal{I}$ can be identified with an element $\phi \in \widehat{C}$. Clearly, $\mathcal{I} = Ker(\phi)$. When thought of as a space of maximal ideals, $\widehat{C}$ is given the *Jacobson topology* (or *hull kernel topology*) producing a space which is homeomorphic to the one constructed by means of the Gel'fand topology. We shall describe the Jacobson topology in detail later .

Example 2.2.1. Let us suppose that the algebra C is generated by N commuting self-adjoint elements $x_1, \ldots, x_N$. Then the structure space $\widehat{C}$ can be identified with a compact subset of $\mathbb{R}^N$ by the map [34],

$$\phi \in \widehat{C} \longrightarrow (\phi(x_1), \ldots, \phi(x_N)) \in \mathbb{R}^N ~, \tag{2.21}$$

and the range of this map is the joint spectrum of $x_1, \ldots, x_N$, namely the set of all N-tuples of eigenvalues corresponding to common eigenvectors.

In general, if $c \in C$, its *Gel'fand transform* $\hat{c}$ is the complex-valued function on $\widehat{C}$, $\hat{c} : \widehat{C} \to \mathbb{C}$, given by

$$\hat{c}(\phi) = \phi(c) ~, \quad \forall~ \phi \in \widehat{C} ~. \tag{2.22}$$

It is clear that $\hat{c}$ is continuous for each c. We thus get the interpretation of elements in C as $\mathbb{C}$-valued continuous functions on $\widehat{C}$. The Gel'fand-Naimark theorem states that all continuous functions on $\widehat{C}$ are of the form (2.22) for some $c \in C$ [43, 65].

Proposition 2.2.1. *Let C be a commutative C^*-algebra. Then, the Gel'fand transform $c \to \hat{c}$ is an isometric $*$-isomorphism of C onto $C(\widehat{C})$; isometric meaning that*

$$||\hat{c}||_\infty = ||c|| ~, \quad \forall~ c \in C ~, \tag{2.23}$$

with $|| \cdot ||_\infty$ the supremum norm on $C(\widehat{C})$ as in (2.9).

---

[3] If there is no unit, one needs to consider ideals which are *regular* (also called *modular*) as well. An ideal $\mathcal{I}$ of a general algebra $\mathcal{A}$ being called regular if there is a unit in $\mathcal{A}$ modulo $\mathcal{I}$, namely an element $u \in \mathcal{A}$ such that $a - au$ and $a - ua$ are in $\mathcal{I}$ for all $a \in \mathcal{A}$ [65]. If $\mathcal{A}$ has a unit, then any ideal is automatically regular.

Suppose now that M is a (locally) compact topological space. As we have seen in Example 2.1.1 of Sect. 2.1, we have a natural C^*-algebra $C(M)$. It is natural to ask what is the relationship between the Gel'fand space $\widehat{C(M)}$ and M itself. It turns out that these two spaces can be identified both setwise and topologically. First of all, each $m \in M$ gives a complex homomorphism $\phi_m \in \widehat{C(M)}$ through the evaluation map,

$$\phi_m : C(M) \to \mathbb{C}, \quad \phi_m(f) = f(m) . \tag{2.24}$$

Let $\mathcal{I}_m$ denote the kernel of ϕ_m, that is the maximal ideal of $C(M)$ consisting of all functions vanishing at m. We have the following [43, 65],

Proposition 2.2.2. *The map ϕ of (2.24) is a homeomorphism of M onto $\widehat{C(M)}$. Equivalently, every maximal ideal of $C(M)$ is of the form $\mathcal{I}_m$ for some $m \in M$.*

The previous two theorems set up a one-to-one correspondence between the *-isomorphism classes of commutative C^*-algebras and the homeomorphism classes of locally compact Hausdorff spaces. Commutative C^*-algebras with unit correspond to compact Hausdorff spaces. In fact, this correspondence is a complete duality between the category of (locally) compact Hausdorff spaces and (proper[4] and) continuous maps and the category of commutative (not necessarily) unital C^*-algebras and *-homomorphisms. Any commutative C^*-algebra can be realized as the C^*-algebra of complex valued functions over a (locally) compact Hausdorff space. Finally, we mention that the space M is metrizable, its topology comes from a metric, if and only if the C^*-algebra is norm separable, meaning that it admits a dense (in the norm) countable subset. Also it is connected if the corresponding algebra has no projectors, i.e. self-adjoint, $p^* = p$, idempotents, $p^2 = p$; this is a consequence of the fact that projectors in a commutative C^*-algebra C correspond to open-closed subsets in its structure space $\widehat{C}$ [33].

2.3 Noncommutative Spaces

The scheme described in the previous Section cannot be directly generalized to a noncommutative C^*-algebra. To show some of the features of the general case, let us consider the simple example (taken from [34]) of the algebra

$$\mathbb{M}_2(\mathbb{C}) = \left\{ \begin{bmatrix} a_{11} & a_{12} \\ a_{21} & a_{22} \end{bmatrix} , \ a_{ij} \in \mathbb{C} \right\} . \tag{2.25}$$

The commutative subalgebra of diagonal matrices

---

[4] We recall that a continuous map between two locally compact Hausdorff spaces $f : X \to Y$ is called *proper* if $f^{-1}(K)$ is a compact subset of X when K is a compact subset of Y.

$$\mathcal{C} = \left\{ \begin{bmatrix} \lambda & 0 \\ 0 & \mu \end{bmatrix} , \; \lambda, \mu \in \mathbb{C} \right\} , \tag{2.26}$$

has a structure space consisting of two points given by the characters

$$\phi_1 \left(\begin{bmatrix} \lambda & 0 \\ 0 & \mu \end{bmatrix} \right) = \lambda , \quad \phi_2 \left(\begin{bmatrix} \lambda & 0 \\ 0 & \mu \end{bmatrix} \right) = \mu . \tag{2.27}$$

These two characters extend as *pure states* (see App. A.2) to the full algebra $\mathbb{M}_2(\mathbb{C})$ as follows,

$$\widetilde{\phi}_i : \mathbb{M}_2(\mathbb{C}) \longrightarrow \mathbb{C} , \; i = 1, 2 ,$$

$$\widetilde{\phi}_1 \left(\begin{bmatrix} a_{11} & a_{12} \\ a_{21} & a_{22} \end{bmatrix} \right) = a_{11} , \quad \widetilde{\phi}_2 \left(\begin{bmatrix} a_{11} & a_{12} \\ a_{21} & a_{22} \end{bmatrix} \right) = a_{22} . \tag{2.28}$$

But now, noncommutativity implies the equivalence of the irreducible representations of $\mathbb{M}_2(\mathbb{C})$ associated, via the Gel'fand-Naimark-Segal construction, with the pure states $\widetilde{\phi}_1$ and $\widetilde{\phi}_2$. In fact, up to equivalence, the algebra $\mathbb{M}_2(\mathbb{C})$ has only one irreducible representation, i.e. the defining two dimensional one.[5] We show this in App. A.2.

For a noncommutative C^*-algebra, there is more than one candidate for the analogue of the topological space M. We shall consider the following ones:

1. The *structure space* of $\mathcal{A}$ or space of all unitary equivalence classes of irreducible *-representations. Such a space is denoted by $\widehat{\mathcal{A}}$.
2. The *primitive spectrum* of $\mathcal{A}$ or the space of kernels of irreducible *-representations. Such a space is denoted by $Prim\mathcal{A}$. Any element of $Prim\mathcal{A}$ is automatically a two-sided *-ideal of $\mathcal{A}$.

While for a commutative C^*-algebra these two spaces agree, this is no longer true for a general C^*-algebra $\mathcal{A}$, not even setwise. For instance, $\widehat{\mathcal{A}}$ may be very complicated while $Prim\mathcal{A}$ consists of a single point. One can define natural topologies on $\widehat{\mathcal{A}}$ and $Prim\mathcal{A}$. We shall describe them in the next Section.

2.3.1 The Jacobson (or Hull-Kernel) Topology

The topology on $Prim\mathcal{A}$ is given by means of a closure operation. Given any subset W of $Prim\mathcal{A}$, the closure $\overline{W}$ of W is by definition the set of all elements in $Prim\mathcal{A}$ containing the intersection $\bigcap W$ of the elements of W, namely

$$\overline{W} =: \{ \mathcal{I} \in Prim\mathcal{A} : \bigcap W \subseteq \mathcal{I} \} . \tag{2.29}$$

For any C^*-algebra $\mathcal{A}$ we have the following,

---

[5] As we mention in App. A.4, $\mathbb{M}_2(\mathbb{C})$ is strongly Morita equivalent to $\mathbb{C}$. In that Appendix we shall also see that two strongly Morita equivalent C^*-algebras have the same space of classes of irreducible representations.

Proposition 2.3.1. *The closure operation (2.29) satisfies the Kuratowski axioms*

K_1 $\bar{\emptyset} = \emptyset$;
K_2 $W \subseteq \overline{W}$, $\forall\, W \in Prim\mathcal{A}$;
K_3 $\overline{\overline{W}} = \overline{W}$, $\forall\, W \in Prim\mathcal{A}$;
K_4 $\overline{W_1 \cup W_2} = \overline{W}_1 \cup \overline{W}_2$, $\forall\, W_1, W_2 \in Prim\mathcal{A}$.

Proof. Property K_1 is immediate since $\bigcap \emptyset$ 'does not exist'. By construction, also K_2 is immediate. Furthermore, $\bigcap \overline{W} = \bigcap W$ from which $\overline{\overline{W}} = \overline{W}$, namely K_3. To prove K_4, first observe that

$$V \subseteq W \;\Longrightarrow\; \left(\bigcap V \right) \supseteq \left(\bigcap W \right) \;\Longrightarrow\; \overline{V} \subseteq \overline{W} . \tag{2.30}$$

From this it follows that $\overline{W}_i \subseteq \overline{W_1 \bigcup W_2}$, $i = 1, 2$ and in turn

$$\overline{W}_1 \cup \overline{W}_2 \subseteq \overline{W_1 \cup W_2} . \tag{2.31}$$

To obtain the opposite inclusion, consider a primitive ideal $\mathcal{I}$ not belonging to $\overline{W}_1 \bigcup \overline{W}_2$. This means that $\bigcap W_1 \not\subseteq \mathcal{I}$ and $\bigcap W_2 \not\subseteq \mathcal{I}$. Thus, if π is a representation of $\mathcal{A}$ with $\mathcal{I} = Ker(\pi)$, there are elements $a \in \bigcap W_1$ and $b \in \bigcap W_2$ such that $\pi(a) \neq 0$ and $\pi(b) \neq 0$. If ξ is any vector in the representation space $\mathcal{H}_\pi$ such that $\pi(a)\xi \neq 0$ then, π being irreducible, $\pi(a)\xi$ is a cyclic vector for π (see App. A.2). This, together with the fact that $\pi(b) \neq 0$, ensures that there is an element $c \in \mathcal{A}$ such that $\pi(b)(\pi(c)\pi(a))\xi \neq 0$ which implies that $bca \neq Ker(\pi) = \mathcal{I}$. But $bca \in (\bigcap W_1) \cap (\bigcap W_2) = \bigcap(W_1 \cup W_2)$. Therefore $\bigcap(W_1 \cup W_2) \not\subseteq \mathcal{I}$; whence $\mathcal{I} \notin \overline{W_1 \cup W_2}$. What we have proven is that $\mathcal{I} \notin \overline{W}_1 \bigcup \overline{W}_2 \Rightarrow \mathcal{I} \notin \overline{W_1 \bigcup W_2}$, which gives the inclusion opposite to (2.31). So K_4 follows.

It also follows that the closure operation (2.29) defines a topology on $Prim\mathcal{A}$, (see App. A.1) which is called *Jacobson topology* or *hull-kernel topology*. The reason for the second name is that $\bigcap W$ is also called the *kernel* of W and then $\overline{W}$ is the *hull* of $\bigcap W$ [65, 43].

To illustrate this topology, we shall give a simple example. Consider the algebra $\mathcal{C}(I)$ of complex-valued continuous functions on an interval I. As we have seen, its structure space $\widehat{\mathcal{C}(I)}$ can be identified with the interval I. For any $a, b \in I$, let W be the subset of $\widehat{\mathcal{C}(I)}$ given by

$$W = \{ \mathcal{I}_x \, , \; x \in \,]a, b[\, \} , \tag{2.32}$$

where $\mathcal{I}_x$ is the maximal ideal of $\mathcal{C}(I)$ consisting of all functions vanishing at the point x,

$$\mathcal{I}_x = \{ f \in \mathcal{C}(I) \mid f(x) = 0 \} . \tag{2.33}$$

The ideal $\mathcal{I}_x$ is the kernel of the evaluation homomorphism as in (2.24). Then

$$\bigcap W = \bigcap_{x \in]a,b[} \mathcal{I}_x = \{f \in \mathcal{C}(I) \mid f(x) = 0 \, , \, \forall \, x \in \,]a,b[\ \} \, , \qquad (2.34)$$

and, the functions being continuous,

$$
\begin{aligned}
\overline{W} &= \{\mathcal{I} \in \widehat{\mathcal{C}} \mid \bigcap W \subset \mathcal{I}\} \\
&= W \bigcup \{\mathcal{I}_a, \mathcal{I}_b\} \\
&= \{\mathcal{I}_x, \ x \in [a,b] \ \} \, , \qquad (2.35)
\end{aligned}
$$

which can be identified with the closure of the interval $]a,b[$.

In general, the space $Prim\mathcal{A}$ has a few properties which are easy to prove. So we simply state them here as propositions [43].

Proposition 2.3.2. *Let W be a subset of $Prim\mathcal{A}$. Then W is closed if and only if W is exactly the set of primitive ideals containing some subset of $\mathcal{A}$.*

Proof. If W is closed then $W = \overline{W}$ and by the very definition (2.29), W is the set of primitive ideals containing $\bigcap W$. Conversely, let $V \subseteq \mathcal{A}$. If W is the set of primitive ideals of $\mathcal{A}$ containing V, then $V \subseteq \bigcap W$ from which $\overline{W} \subset W$, and, in turn, $\overline{W} = W$.

Proposition 2.3.3. *There is a bijective correspondence between closed subsets W of $Prim\mathcal{A}$ and (norm-closed two sided) ideals $\mathcal{J}_W$ of $\mathcal{A}$. The correspondence is given by*

$$W = \{\mathcal{I} \in Prim\mathcal{A} \mid \mathcal{J}_W \subseteq \mathcal{I}\} \, . \qquad (2.36)$$

Proof. If W is closed then $W = \overline{W}$ and by the definition (2.29), $\mathcal{J}_W$ is just the ideal $\bigcap W$. Conversely, from the previous proposition, W defined as in (2.36) is closed.

Proposition 2.3.4. *Let W be a subset of $Prim\mathcal{A}$. Then W is closed if and only if $\mathcal{I} \in W$ and $\mathcal{I} \subseteq \mathcal{J} \ \Rightarrow \ \mathcal{J} \in W$.*

Proof. If W is closed then $W = \overline{W}$ and by the definition (2.29), $\mathcal{I} \in W$ and $\mathcal{I} \subseteq \mathcal{J}$ implies that $\mathcal{J} \in W$. The converse implication is also evident by the previous Proposition.

Proposition 2.3.5. *The space $Prim\mathcal{A}$ is a T_0-space.*[6]

Proof. Suppose $\mathcal{I}_1$ and $\mathcal{I}_2$ are two distinct points of $Prim\mathcal{A}$ so that, say, $\mathcal{I}_1 \not\subset \mathcal{I}_2$. Then the set W of those $\mathcal{I} \in Prim\mathcal{A}$ which contain $\mathcal{I}_1$ is a closed subset (by 2.3.2), such that $\mathcal{I}_1 \in W$ and $\mathcal{I}_2 \notin W$. The complement W^c of W is an open set containing $\mathcal{I}_2$ and not $\mathcal{I}_1$.

---

[6] We recall that a topological space is called T_0 if for any two distinct points of the space there is an open neighborhood of one of the points which does not contain the other [59].

Proposition 2.3.6. *Let $\mathcal{I} \in Prim\mathcal{A}$. Then the point $\{\mathcal{I}\}$ is closed in $Prim\mathcal{A}$ if and only if $\mathcal{I}$ is maximal among primitive ideals.*

Proof. Indeed, the closure of $\{\mathcal{I}\}$ is just the set of primitive ideals of $\mathcal{A}$ containing $\mathcal{I}$.

In general, $Prim\mathcal{A}$ is not a T_1-space[7] and will be so if and only if all primitive ideals in $\mathcal{A}$ are also maximal. The notion of primitive ideal is more general than the one of maximal ideal. For a commutative C^*-algebra an ideal is primitive if and only if it is maximal. For a general $\mathcal{A}$ (with unit), while a maximal ideal is primitive, the converse need not be true [43].

Let us now consider the structure space $\widehat{\mathcal{A}}$. Now, there is a canonical surjection

$$\widehat{\mathcal{A}} \longrightarrow Prim\mathcal{A} , \quad \pi \mapsto ker(\pi) . \tag{2.37}$$

The inverse image under this map, of the Jacobson topology on $Prim\mathcal{A}$ is a topology for $\widehat{\mathcal{A}}$. In this topology, a subset $S \subset \widehat{\mathcal{A}}$ is open if and only if it is of the form $\{\pi \in \widehat{\mathcal{A}} \mid ker(\pi) \in W\}$ for some subset $W \subset Prim\mathcal{A}$ which is open in the (Jacobson) topology of $Prim\mathcal{A}$. The resulting topological space is still called the structure space. There is another natural topology on the space $\widehat{\mathcal{A}}$ called the *regional topology*. For a C^*-algebra $\mathcal{A}$, the regional and the pullback of the Jacobson topology on $\widehat{\mathcal{A}}$ coincide, [65, page 563].

Proposition 2.3.7. *Let $\mathcal{A}$ be a C^*-algebra. The following conditions are equivalent*

(i) $\widehat{\mathcal{A}}$ is a T_0 space;
(ii) Two irreducible representations of $\widehat{\mathcal{A}}$ with the same kernel are equivalent;
(iii) The canonical map $\widehat{\mathcal{A}} \to Prim\mathcal{A}$ is a homeomorphism.

Proof. By construction, a subset $S \in \widehat{\mathcal{A}}$ will be closed if and only if it is of the form $\{\pi \in \widehat{\mathcal{A}} : ker(\pi) \in W\}$ for some W closed in $Prim\mathcal{A}$. As a consequence, given any two (classes of) representations $\pi_1, \pi_2 \in \widehat{\mathcal{A}}$, the representation π_1 will be in the closure of π_2 if and only if $ker(\pi_1)$ is in the closure of $ker(\pi_2)$, or, by Prop.2.3.2 if and only if $ker(\pi_2) \subset ker(\pi_1)$. In turn, π_1 and π_2 agree in the closure of the other if and only if $ker(\pi_2) = ker(\pi_1)$. Therefore, π_1 and π_2 will not be distinguished by the topology of $\widehat{\mathcal{A}}$ if and only if they have the same kernel. On the other side, if $\widehat{\mathcal{A}}$ is T_0 one is able to distinguish points. It follows that (i) implies that two representations with the same kernel must be equivalent so as to correspond to the same point of $\widehat{\mathcal{A}}$, namely (ii). The other implications are obvious.

---

[7] We recall that a topological space is called T_1 if any point of the space is closed [59].

We recall that a (not necessarily Hausdorff) topological space S is called locally compact if any point of S has at least one compact neighborhood. A compact space is automatically locally compact. If S is a locally compact space which is also Hausdorff, then the family of closed compact neighborhoods of any point is a base for its neighborhood system. With respect to compactness, the structure space of a noncommutative C^*-algebra behaves as in the commutative situation [65, page 576]; we have

Proposition 2.3.8. *If A is a C^*-algebra, then $\widehat{A}$ is locally compact. Likewise, $\mathrm{Prim}A$ is locally compact. If A has a unit, then both $\widehat{A}$ and $\mathrm{Prim}A$ are compact.*

Notice that in general, $\widehat{A}$ being compact does not imply that A has a unit. For instance, the algebra $\mathcal{K}(\mathcal{H})$ of compact operators on an infinite dimensional Hilbert space $\mathcal{H}$ has no unit but its structure space has only one point (see the next Section).

2.4 Compact Operators

We recall [117] that an operator on the Hilbert space $\mathcal{H}$ is said to be of *finite rank* if the orthogonal complement of its null space is finite dimensional. Essentially, we may think of such an operator as a finite dimensional matrix even if the Hilbert space is infinite dimensional.

Definition 2.4.1. *An operator T on $\mathcal{H}$ is said to be compact if it can be approximated in norm by finite rank operators.*

An equivalent way to characterize a compact operator T is by stating that

$$\forall\, \varepsilon > 0\, ,\quad \exists\ \text{a finite dimensional subspace } E \subset \mathcal{H} \mid \|T|_{E^\perp}\| < \varepsilon\, . \quad (2.38)$$

Here the orthogonal subspace $E^\perp$ is of finite codimension in $\mathcal{H}$. The set $\mathcal{K}(\mathcal{H})$ of all compact operators T on the Hilbert space $\mathcal{H}$ is the largest two-sided ideal in the C^*-algebra $\mathcal{B}(\mathcal{H})$ of all bounded operators. In fact, it is the only norm closed and two-sided ideal when $\mathcal{H}$ is separable; and it is essential [65]. It is also a C^*-algebra without a unit, since the operator $\mathbb{I}$ on an infinite dimensional Hilbert space is not compact. The defining representation of $\mathcal{K}(\mathcal{H})$ by itself is irreducible [65] and it is the *only* irreducible representation of $\mathcal{K}(\mathcal{H})$ up to equivalence.[8]

There is a special class of C^*-algebras which have been used in a scheme of approximation by means of topological lattices [6, 7, 11]; they are *postliminal* algebras. For these algebras, a relevant rôle is again played by the compact

---

[8] If $\mathcal{H}$ is finite dimensional, $\mathcal{H} = \mathbb{C}^n$ say, then $\mathcal{B}(\mathbb{C}^n) = \mathcal{K}(\mathbb{C}^n) = \mathbb{M}_n(\mathbb{C})$, the algebra of $n \times n$ matrices with complex entries. Such an algebra has only one irreducible representation (as an algebra), namely the defining one.

operators. Before we give the appropriate definitions, we state another result which shows the relevance of compact operators in the analysis of irreducibility of representations of a general C^*-algebra and which is a consequence of the fact that $\mathcal{K}(\mathcal{H})$ is the largest two-sided ideal in $\mathcal{B}(\mathcal{H})$ [108].

Proposition 2.4.1. *Let $\mathcal{A}$ be a C^*-algebra acting irreducibly on a Hilbert space $\mathcal{H}$ and having non-zero intersection with $\mathcal{K}(\mathcal{H})$. Then $\mathcal{K}(\mathcal{H}) \subseteq \mathcal{A}$.*

Definition 2.4.2. *A C^*-algebra $\mathcal{A}$ is said to be liminal if for every irreducible representation $(\mathcal{H}, \pi)$ of $\mathcal{A}$ one has that $\pi(\mathcal{A}) = \mathcal{K}(\mathcal{H})$ (or equivalently, from Prop. 2.4.1, $\pi(\mathcal{A}) \subset \mathcal{K}(\mathcal{H})$).*

So, the algebra $\mathcal{A}$ is liminal if it is mapped to the algebra of compact operators under any irreducible representation. Furthermore, if $\mathcal{A}$ is a liminal algebra, then one can prove that each primitive ideal of $\mathcal{A}$ is automatically a maximal closed two-sided ideal of $\mathcal{A}$. As a consequence, all points of $Prim\mathcal{A}$ are closed and $Prim\mathcal{A}$ is a T_1 space. In particular, every commutative C^*-algebra is liminal [108, 43].

Definition 2.4.3. *A C^*-algebra $\mathcal{A}$ is said to be postliminal if for every irreducible representation $(\mathcal{H}, \pi)$ of $\mathcal{A}$ one has that $\mathcal{K}(\mathcal{H}) \subseteq \pi(\mathcal{A})$ (or equivalently, from Prop. 2.4.1, $\pi(\mathcal{A}) \cap \mathcal{K}(\mathcal{H}) \neq 0$).*

Every liminal C^*-algebra is postliminal but the converse is not true. Postliminal algebras have the remarkable property that their irreducible representations are completely characterized by the kernels: if $(\mathcal{H}_1, \pi_1)$ and $(\mathcal{H}_2, \pi_2)$ are two irreducible representations with the same kernel, then π_1 and π_2 are equivalent [108, 43]. From Prop. (2.3.7), the spaces $\widehat{\mathcal{A}}$ and $Prim\mathcal{A}$ are then homeomorphic.

2.5 Real Algebras and Jordan Algebras

From what we have said up to now, it should be clear that a (complex) C^*-algebra replaces the algebra of complex valued continuous functions on a topological space. It is natural then, to look for suitable replacements of the algebra of real valued functions. Well, it has been suggested (see for instance [49]) that rather than an associative real algebra, in the spirit of quantum mechanics, one should consider the Jordan algebra $\mathcal{A}^{sa}$ of self-adjoint elements of a complex C^*-algebra $\mathcal{A}$,

$$\mathcal{A}^{sa} =: \{a \in \mathcal{A} \mid a^* = a\} . \tag{2.39}$$

Now, $\mathcal{A}^{sa}$ has a natural product

$$a \circ b =: \frac{1}{2}(ab + ba) , \quad \forall \, a, b \in \mathcal{A}^{sa} , \tag{2.40}$$

which is *commutative* but *non associative*, as can be easily checked. Instead, commutativity gives a weak form of associativity which goes under the name of the *Jordan algebra identity*,

$$a^2 \circ (a \circ b) = a \circ (a^2 \circ b) , \quad \forall \, a, b \in \mathcal{A}^{sa} . \tag{2.41}$$

The product (2.40) is the only one which preserves $\mathcal{A}^{sa}$.

We refer to [129] for an introduction to Jordan algebras and to [136] for some work on *nonassociative geometry* and its use in gauge theories.

3. Projective Systems
of Noncommutative Lattices

The idea of a 'discrete substratum' underpinning the 'continuum' is somewhat spread among physicists. With particular emphasis this idea has been pushed by R. Sorkin who, in [121], assumes that the substratum be a *finitary* (see later) topological space which maintains some of the topological information of the continuum. It turns out that the finitary topology can be equivalently described in terms of a partial order. This partial order has been alternatively interpreted as determining the causal structure in the approach to quantum gravity of [14]. Recently, finitary topological spaces have been interpreted as noncommutative lattices and noncommutative geometry has been used to construct quantum mechanical and field theoretical models, notably lattice field theory models, on them [6, 7].

Given a suitable covering of a topological space M, by identifying any two points of M which cannot be 'distinguished' by the sets in the covering, one constructs a lattice with a finite (or in general a countable) number of points. Such a lattice, with the quotient topology, becomes a T_0-space which turns out to be the structure space (or equivalently, the space of primitive ideals) of a postliminar approximately finite dimensional (AF) algebra. Therefore the lattice is truly a noncommutative space. In this Chapter we shall describe noncommutative lattices in some detail while in Chap. 11 we shall illustrate some of their applications in physics.

3.1 The Topological Approximation

The approximation scheme we are going to describe has really a deep physical flavour. To get a taste of the general situation, let us consider the following simple example. Let us suppose we are about to measure the position of a particle which moves on a circle, of radius one say, $S^1 = \{0 \leq \varphi \leq 2\pi, \bmod 2\pi\}$. Our *'detectors'* will be taken to be (possibly overlapping) open subsets of S^1 with some mechanism which switches on the detector when the particle is in the corresponding open set. The number of detectors must be clearly limited and we take them to consist of the following three open subsets whose union covers S^1,

$$U_1 = \{-\tfrac{1}{3}\pi < \varphi < \tfrac{2}{3}\pi\},$$

$$U_2 = \{\tfrac{1}{3}\pi < \varphi < \tfrac{4}{3}\pi\}, \qquad\qquad (3.1)$$

$$U_3 = \{\pi < \varphi < 2\pi\}.$$

Now, if two detectors, U_1 and U_2 say, are on, we will know that the particle is in the intersection $U_1 \cap U_2$ although we will be unable to distinguish any two points in this intersection. The same will be true for the other two intersections. Furthermore, if only one detector, U_1 say, is on, we can infer the presence of the particle in the *closed* subset of S^1 given by $U_1 \backslash \{U_1 \cap U_2 \bigcup U_1 \cap U_3\}$ but again we will be unable to distinguish any two points in this closed set. The same will be true for the other two closed sets of similar type. Summing up, if we have only the three detectors (3.1), we are forced to identify the points which cannot be distinguished and S^1 will be represented by a collection of six points $P = \{\alpha, \beta, \gamma, a, b, c\}$ which correspond to the following identifications

$$U_1 \cap U_3 = \{\tfrac{5}{3}\pi < \varphi < 2\pi\} \qquad\qquad \rightarrow \quad \alpha,$$

$$U_1 \cap U_2 = \{\tfrac{1}{3}\pi < \varphi < \tfrac{2}{3}\pi\} \qquad\qquad \rightarrow \quad \beta,$$

$$U_2 \cap U_3 = \{\pi < \varphi < \tfrac{4}{3}\pi\} \qquad\qquad \rightarrow \quad \gamma,$$

$$\qquad\qquad\qquad\qquad\qquad\qquad\qquad\qquad\qquad\qquad (3.2)$$

$$U_1 \backslash \{U_1 \cap U_2 \bigcup U_1 \cap U_3\} = \{0 \le \varphi \le \tfrac{1}{3}\pi\} \quad \rightarrow \quad a,$$

$$U_2 \backslash \{U_2 \cap U_1 \bigcup U_2 \cap U_3\} = \{\tfrac{2}{3}\pi \le \varphi \le \pi\} \quad \rightarrow \quad b,$$

$$U_3 \backslash \{U_3 \cap U_2 \bigcup U_3 \cap U_1\} = \{\tfrac{4}{3}\pi \le \varphi \le \tfrac{5}{3}\pi\} \quad \rightarrow \quad c.$$

We can push things a bit further and keep track of the kind of set from which a point in P comes by declaring the point to be open (respectively closed) if the subset of S^1 from which it comes is open (respectively closed). This is equivalently achieved by endowing the space P with a topology a basis of which is given by the following open (by definition) sets,

$$\{\alpha\}, \ \{\beta\}, \ \{\gamma\},$$
$$\{\alpha, a, \beta\}, \ \{\beta, b, \gamma\}, \ \{\alpha, c, \gamma\} \ . \qquad (3.3)$$

The corresponding topology on the quotient space P is nothing but the quotient topology of the one on S^1 generated by the three open sets U_1, U_2, U_3, by the quotient map (3.2).

In general, let us suppose that we have a topological space M together with an open covering $\mathcal{U} = \{U_\lambda\}$ which is also a topology for M, so that $\mathcal{U}$ is closed under arbitrary unions and finite intersections (see App. A.1). We define an equivalence relation among points of M by declaring that any two points $x, y \in M$ are equivalent if every open set U_λ containing either x or y contains the other too,

$$x \sim y \quad \text{if and only if} \quad x \in U_\lambda \Leftrightarrow y \in U_\lambda , \quad \forall \ U_\lambda \in \mathcal{U} . \tag{3.4}$$

Thus, two points of M are identified if they cannot be distinguished by any 'detector' in the collection $\mathcal{U}$.

The space $P_{\mathcal{U}}(M) =: M/\sim$ of equivalence classes is then given the quotient topology. If $\pi : M \to P_{\mathcal{U}}(M)$ is the natural projection, a set $U \subset P_{\mathcal{U}}(M)$ is declared to be open if and only if $\pi^{-1}(U)$ is open in the topology of M given by $\mathcal{U}$. The quotient topology is the finest one making π continuous. When M is compact, the covering $\mathcal{U}$ can be taken to be finite so that $P_{\mathcal{U}}(M)$ will consist of a finite number of points. If M is only locally compact the covering can be taken to be locally finite and each point has a neighborhood intersected by only finitely many U_λ' s. Then the space $P_{\mathcal{U}}(M)$ will consist of a countable number of points; in the terminology of [121] $P_{\mathcal{U}}(M)$ would be a *finitary* approximation of M. If $P_{\mathcal{U}}(M)$ has N points we shall also denote it by $P_N(M)$.[1] For example, the finite space given by (3.2) is $P_6(S^1)$.

In general, $P_{\mathcal{U}}(M)$ is not Hausdorff: from (3.3) it is evident that in $P_6(S^1)$, for instance, we cannot isolate the point a from α by using open sets. It is not even a T_1-space; again, in $P_6(S^1)$ only the points a, b and c are closed while the points α, β and γ are open. In general there will be points which are neither closed nor open. It can be shown, however, that $P_{\mathcal{U}}(M)$ is always a T_0-space, being, indeed, the T_0-quotient of M with respect to the topology $\mathcal{U}$ [121].

3.2 Order and Topology

What we shall show next is how the topology of any finitary T_0 topological space P can be given equivalently by means of a partial order which makes P a *partially ordered set* (or *poset* for short) [121]. Consider first the case when P is finite. Then, the collection τ of open sets (the topology on P) will be closed under arbitrary unions and arbitrary intersections. As a consequence, for any point $x \in P$, the intersection of all open sets containing it,

$$\Lambda(x) =: \bigcap \{ U \in \tau \mid x \in U \} , \tag{3.5}$$

will be the smallest open neighborhood containing the point. A relation $\preceq$ is then defined on P by

$$x \preceq y \quad \Leftrightarrow \quad \Lambda(x) \subseteq \Lambda(y) , \quad \forall \ x, y \in P . \tag{3.6}$$

---

[1] In fact, this notation is incomplete since it does not keep track of the finite topology given on the set of N points. However, at least for the examples considered in these notes, the topology will always be given explicitly.

Now, $x \in \Lambda(x)$ always, so that the previous definition is equivalent to

$$x \preceq y \iff x \in \Lambda(y) , \tag{3.7}$$

which can also be stated as saying that

$$x \preceq y \iff \text{every open set containing } y \text{ also contains } x , \tag{3.8}$$

or, in turn, that

$$x \preceq y \iff y \in \overline{\{x\}} , \tag{3.9}$$

with $\overline{\{x\}}$ the closure of the one point set $\{x\}$.[2]
From (3.6) it is clear that the relation $\preceq$ is reflexive and transitive,

$$\begin{aligned} &x \preceq x, \\ &x \preceq y , \ y \preceq z \implies x \preceq z . \end{aligned} \tag{3.10}$$

Furthermore, since P is a T_0-space, for any two distinct points $x, y \in P$, there is at least one open set containing x, say, and not y. This, together with (3.8), implies that the relation $\preceq$ is symmetric as well,

$$x \preceq y , \ y \preceq x \implies x = y . \tag{3.11}$$

Summing up, we see that a T_0 topology on a finite space P determines a reflexive, antisymmetric and transitive relation, namely a *partial order* on P which makes the latter a *partially ordered set* (*poset*). Conversely, given a partial order $\preceq$ on the set P, one produces a topology on P by taking as a basis for it the finite collection of 'open' sets defined by

$$\Lambda(x) =: \{y \in P \mid y \preceq x\} , \ \forall \, x \in P . \tag{3.12}$$

Thus, a subset $W \subset P$ will be open if and only if it is the union of sets of the form (3.12), that is, if and only if $x \in W$ and $y \preceq x \implies y \in W$. Indeed, the smallest open set containing W is given by

$$\Lambda(W) = \bigcup_{x \in W} \Lambda(x) , \tag{3.13}$$

and W is open if and only if $W = \Lambda(W)$.
The resulting topological space is clearly T_0 by the antisymmetry of the order relation.

It is easy to express the closure operation in terms of the partial order. From (3.9), the closure $V(x) = \overline{\{x\}}$, of the one point set $\{x\}$ is given by

$$V(x) =: \{y \in P \mid x \preceq y\} , \ \forall \, x \in P . \tag{3.14}$$

---

[2] Another equivalent definition can be given by saying that $x \preceq y$ if and only if the constant sequence $(x, x, x, \cdots)$ converges to y. It is worth noticing that in a T_0-space the limit of a sequence need not be unique so that the constant sequence $(x, x, x, \cdots)$ may converge to more than one point.

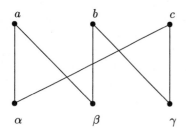 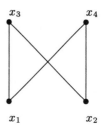

Fig. 3.1. The Hasse diagrams for $P_6(S^1)$ and for $P_4(S^1)$

A subset $W \subset P$ will be closed if and only if $x \in W$ and $x \preceq y \Rightarrow y \in W$. Indeed, the closure of W is given by

$$V(W) = \bigcup_{x \in W} V(x) , \qquad (3.15)$$

and W is closed if and only if $W = V(W)$.

If one relaxes the condition of finiteness of the space P, there is still an equivalence between topology and partial order for any T_0 topological space which has the additional property that every intersection of open sets is an open set (or equivalently, that every union of closed sets is a closed set), so that the sets (3.5) are all open and provide a basis for the topology [4, 21]. This would be the case if P were a finitary approximation of a (locally compact) topological space M, obtained then from a locally finite covering of M.[3]

Given two posets P, Q, it is clear that a map $f : P \to Q$ will be continuous if and only if it is *order preserving*, i.e., if and only if from $x \preceq_P y$ it follows that $f(x) \preceq_Q f(y)$; indeed, f is continuous if and only if it preserves convergence of sequences.

In the sequel, $x \prec y$ will indicate that x precedes y while $x \neq y$.

A pictorial representation of the topology of a poset is obtained by constructing the associated *Hasse diagram*: one arranges the points of the poset at different levels and connects them by using the following rules :

1. if $x \prec y$, then x is at a lower level than y;
2. if $x \prec y$ and there is no z such that $x \prec z \prec y$, then x is at the level immediately below y and these two points are connected by a link.

Figure 3.1 shows the Hasse diagram for $P_6(S^1)$ whose basis of open sets is in (3.3) and for $P_4(S^1)$. For the former, the partial order reads $\alpha \prec a, \ \alpha \prec$

---

[3] In fact, Sorkin [121] regards as finitary only those posets P for which the sets $\Lambda(x)$ and $V(x)$ defined in (3.13) and (3.14) respectively, are all finite. This would be the case if the poset were derived from a locally compact topological space with a locally finite covering consisting of bounded open sets.

c, $\beta \prec a$, $\beta \prec b$, $\gamma \prec b$, $\gamma \prec c$. The latter is a four point approximation of S^1 obtained from a covering consisting of two intersecting open sets. The partial order reads $x_1 \prec x_3$, $x_1 \prec x_4$, $x_2 \prec x_3$, $x_2 \prec x_4$.

In Fig. 3.1, (and in general, in any Hasse diagram) the smallest open set containing any point x consists of all points which are below the given one, x, and can be connected to it by a series of links. For instance, for $P_4(S^1)$, we get the following collection for the minimal open sets,

$$\Lambda(x_1) = \{x_1\}, \qquad \Lambda(x_2) = \{x_2\},$$

$$\Lambda(x_3) = \{x_1, x_2, x_3\}, \quad \Lambda(x_4) = \{x_1, x_2, x_4\}, \tag{3.16}$$

which are a basis for the topology of $P_4(S^1)$.

The generic finitary poset $P(\mathbb{R})$ associated with the real line $\mathbb{R}$ is shown in Fig. 3.2. The corresponding projection $\pi : \mathbb{R} \to P(\mathbb{R})$ is given by

$$U_i \cap U_{i+1} \longrightarrow x_i, \quad i \in \mathbb{Z},$$
$$U_{i+1} \setminus \{U_i \cap U_{i+1} \cup U_{i+1} \cap U_{i+2}\} \longrightarrow y_i, \quad i \in \mathbb{Z}. \tag{3.17}$$

A basis for the quotient topology is provided by the collection of all open sets of the form

$$\Lambda(x_i) = \{x_i\}, \quad \Lambda(y_i) = \{x_i, y_i, x_{i+1}\}, \quad i \in \mathbb{Z}. \tag{3.18}$$

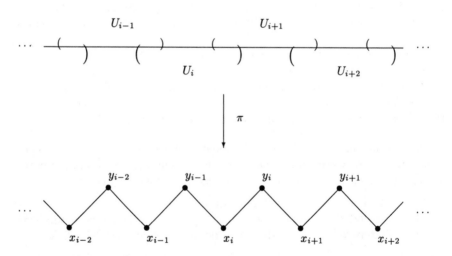

Fig. 3.2. The finitary poset of the line $\mathbb{R}$

Figure 3.3 shows the Hasse diagram for the six-point poset $P_6(S^2)$ of the two dimensional sphere, coming from a covering with four open sets, which was derived in [121]. A basis for its topology is given by

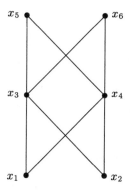

Fig. 3.3. The Hasse diagram for the poset $P_6(S^2)$

$$\Lambda(x_1) = \{x_1\} , \qquad\qquad \Lambda(x_2) = \{x_2\} ,$$

$$\Lambda(x_3) = \{x_1, x_2, x_3\} , \qquad \Lambda(x_4) = \{x_1, x_2, x_4\} , \qquad (3.19)$$

$$\Lambda(x_5) = \{x_1, x_2, x_3, x_4, x_5\} , \quad \Lambda(x_6) = \{x_1, x_2, x_3, x_4, x_6\} .$$

Now, the top two points are closed, the bottom two points are open and the intermediate ones are neither closed nor open.

As alluded to before, posets retain some of the topological information of the space they approximate. For example, one can prove that for the first homotopy group, $\pi_1(P_N(S^1)) = \mathbb{Z} = \pi(S^1)$ whenever $N \geq 4$ [121]. Consider the case $N = 4$. Elements of $\pi_1(P_4(S^1))$ are homotopy classes of continuous maps $\sigma : [0,1] \to P_4(S^1)$, such that $\sigma(0) = \sigma(1)$. With a any real number in the open interval $]0,1[$, consider the map

$$\sigma(t) = \begin{cases} x_3 & if \quad t = 0 \\ x_2 & if \quad 0 < t < a \\ x_4 & if \quad t = a \\ x_1 & if \quad a < t < 1 \\ x_3 & if \quad t = 1 \end{cases} . \qquad (3.20)$$

Figure 3.4 shows this map for $a = 1/2$; the map can be seen to 'wind once around' $P_4(S^1)$. Furthermore, the map σ in (3.20) is manifestly continuous, being constructed in such a way that closed (respectively open) points of $P_4(S^1)$ are the images of closed (respectively open) sets of the interval $[0,1]$. Hence it is automatic that the inverse image of an open set in $P_4(S^1)$ is open in $[0,1]$. A bit of extra analysis shows that σ is not contractible to the constant map: Any such contractible map being one that skips at least one of the points of $P_4(S^1)$, like the following one

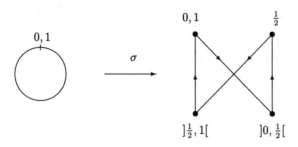

Fig. 3.4. A representative of the generator of the homotopy group $\pi_1(P_4(S^1))$

$$\sigma_0(t) = \left\{ \begin{array}{llll} x_3 & if & t = 0 \\ x_2 & if & 0 < t < a \\ x_4 & if & t = a \\ x_2 & if & a < t < 1 \\ x_3 & if & t = 1 \end{array} \right. \tag{3.21}$$

This is shown in Fig. 3.5 for the value $a = 1/2$. Indeed, the not contractible

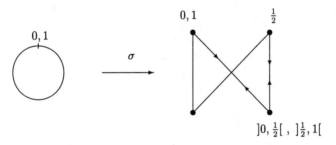

Fig. 3.5. A representative of the trivial class in the homotopy group $\pi_1(P_4(S^1))$

map in (3.20) is a generator of the group $\pi_1(P_4(S^1))$ which can, therefore, be identified with the group of integers $\mathbb{Z}$.

Finally, we mention the notion of a Cartesian product of posets. If P and Q are posets, their *Cartesian product* is the poset $P \times Q$ on the set $\{(x, y) \mid x \in P, \ y \in Q\}$ such that $(x, y) \preceq (x', y')$ in $P \times Q$ if $x \preceq x'$ in P and $y \preceq y'$ in Q. To draw the Hasse diagram of $P \times Q$, one draws the diagram of P, replaces each element x of P with a copy Q_x of Q and connects corresponding elements of Q_x and Q_y (by identifying $Q_x \simeq Q_y$) if x and y are connected in the diagram of P. Figure 3.6 shows the Hasse diagram of a poset $P_{16}(S^1 \times S^1)$ obtained as $P_4(S^1) \times P_4(S^1)$.

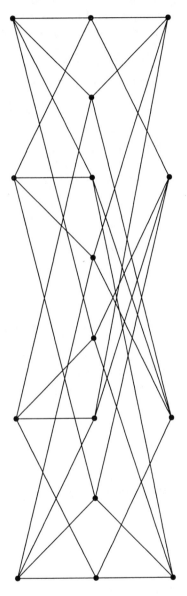

Fig. 3.6. The Hasse diagram for the poset $P_{16}(S^1 \times S^1) = P_4(S^1) \times P_4(S^1)$ (the diagram has been rotated; the levels are counted horizontally)

3.3 How to Recover the Space Being Approximated

We shall now briefly describe how the topological space being approximated can be recovered 'in the limit' by considering a sequence of finer and finer coverings, the appropriate framework being that of projective (or inverse) systems of topological spaces [121].

Well, let us suppose we have a topological space M together with a sequence $\{\mathcal{U}_n\}_{n \in \mathbb{N}}$ of finer and finer coverings, that is of coverings such that

$$\mathcal{U}_i \subseteq \tau(\mathcal{U}_{i+1}) , \tag{3.22}$$

where $\tau(\mathcal{U})$ is the topology generated by the covering $\mathcal{U}$.[4] Here we are relaxing the harmless assumption made in Sect. 3.1 that each $\mathcal{U}$ is already a subtopology, namely that $\mathcal{U} = \tau(\mathcal{U})$.

In Sect. 3.1 we have associated with each covering $\mathcal{U}_i$ a T_0-topological space P_i and a continuous surjection

$$\pi_i : M \to P_i . \tag{3.23}$$

We now construct a *projective system of spaces* P_i together with *continuous maps*

$$\pi_{ij} : P_j \to P_i , \tag{3.24}$$

defined whenever $i \leq j$ and such that

$$\pi_i = \pi_{ij} \circ \pi_j . \tag{3.25}$$

These maps are uniquely defined by the fact that the spaces P_i's are T_0 and that the map π_i is continuous with respect to $\tau(\mathcal{U}_j)$ whenever $i \leq j$. Indeed, if U is open in P_i, then $\pi_i^{(-1)}(U)$ is open in the $\mathcal{U}_i$-topology by definition, thus it is also open in the finer $\mathcal{U}_j$-topology and π_i is continuous in $\tau(\mathcal{U}_j)$. Furthermore, uniqueness also implies the compatibility conditions

$$\pi_{ij} \circ \pi_{jk} = \pi_{ik} , \tag{3.26}$$

whenever $i \leq j \leq k$.[5] Notice that from the surjectivity of the maps π_i's and the relation (3.25), it follows that all maps π_{ij} are surjective.

The projective system of topological spaces together with continuous maps $\{P_i, \pi_{ij}\}_{i,j \in \mathbb{N}}$ has a unique *projective limit*, namely a topological space P_∞, together with continuous maps

---

[4] For more general situations, such as the system of all finite open covers of M, this is not enough and one needs to consider a *directed* collection $\{\mathcal{U}_i\}_{i \in \Lambda}$ of open covers of M. Here directed just means that for any two coverings $\mathcal{U}_1$ and $\mathcal{U}_2$, there exists a third cover $\mathcal{U}_3$ such that $\mathcal{U}_1, \mathcal{U}_2 \subseteq \tau(\mathcal{U}_3)$. The construction of the remaining part of this Section applies to this more general situation if one defines a partial order on the 'set of indices' Λ by declaring that $i \leq k \Leftrightarrow \mathcal{U}_i \subseteq \tau(\mathcal{U}_j)$ for any $j, k \in \Lambda$.

[5] Indeed, the map π_{ij} is the solution (by definition it is then unique) of a universal mapping problem for maps relating T_0-spaces [121].

$$\pi_{i\infty} : P_\infty \to P_i , \qquad (3.27)$$

such that

$$\pi_{ij} \circ \pi_{j\infty} = \pi_{i\infty} , \qquad (3.28)$$

whenever $i \leq j$. The space P_∞ and the maps π_{ij} can be constructed explicitly. An element $x \in P_\infty$ is an arbitrary coherent sequence of elements $x_i \in P_i$,

$$x = (x_i)_{i\in\mathbb{N}} , \ x_i \in P_i \mid \exists \, N_0 \quad \text{s.t.} \quad x_i = \pi_{i,i+1}(x_{i+1}) , \ \forall \, i \geq N_0 . \quad (3.29)$$

As for the map $\pi_{i\infty}$, it is simply defined by

$$\pi_{i\infty}(x) = x_i . \qquad (3.30)$$

The space P_∞ is made into a T_0 topological space by endowing it with the weakest topology making all maps $\pi_{i\infty}$ continuous: a basis for it is given by the sets $\pi_{i\infty}^{(-1)}(U)$, for all open sets $U \subset P_i$. The projective system and its limit are depicted in Fig. 3.7.

It turns out that the limit space P_∞ is *bigger* than the starting space M and that the latter is contained as a dense subspace. Furthermore, M can be characterized as the set of all *closed* points of $P_{i\infty}$. Let us first observe that we also get a unique (by universality) continuous map

$$\pi_\infty : M \to P_\infty , \qquad (3.31)$$

which satisfies

$$\pi_i = \pi_{i\infty} \circ \pi_\infty , \quad \forall \, i \in \mathbb{N} . \qquad (3.32)$$

The map π_∞ is the 'limit' of the maps π_i. However, while the latter are surjective, under mild hypothesis the former turns out to be *injective*. We have, indeed, the following two propositions [121].

Proposition 3.3.1. *The image $\pi_\infty(M)$ is dense in P_∞.*

Proof. If $U \subset P_\infty$ is any nonempty open set, by the definition of the topology of P_∞, U is the union of sets of the form $\pi_{i\infty}^{(-1)}(U_i)$, with U_i open in P_i. Choose $x_i \in U_i$. Since π_i is surjective, there is at least a point $m \in M$, for which $\pi_i(m) = x_i$ and let $\pi_\infty(m) = x$. Then $\pi_{i\infty}(m) = \pi_{i\infty}(\pi_\infty)(m) = x_j$, from which $x \in \pi_{i\infty}^{(-1)}(x_i) \subset \pi_{i\infty}^{(-1)}(U_i) \subset U$. This proves that $\pi_\infty(M) \cap W \neq \emptyset$, which establishes that $\pi_\infty(M)$ is dense. $\quad\blacksquare$

Proposition 3.3.2. *Let M be T_0 and the collection $\{\mathcal{U}_i\}$ of coverings be such that for every $m \in M$ and every neighborhood $N \ni m$, there exists an index i and an element $U \in \tau(\mathcal{U}_i)$ such that $m \in U \subset N$. Then, the map π_∞ is injective.*

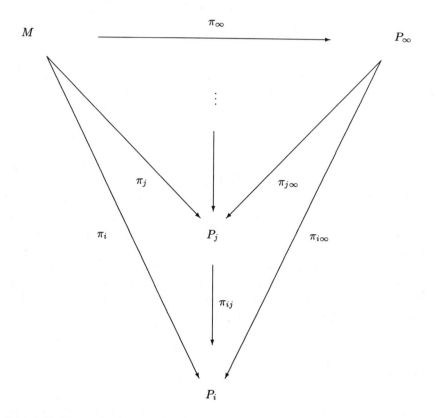

Fig. 3.7. The projective system of topological spaces with continuous maps which approximates the space M

Proof. If m_1, m_2 are two distinct points of M, since the latter is T_0, there is an open set V containing m_1 (say) and not m_2. By hypothesis, there exists an index i and an open $U \in \tau(\mathcal{U}_i)$ such that $m_1 \in U \subset V$. Therefore $\tau(\mathcal{U}_i)$ distinguishes m_1 from m_2. Since P_i is the corresponding T_0 quotient, $\pi_i(m_1) \neq \pi_i(m_2)$. Then $\pi_{i\infty}(\pi_\infty(m_1)) \neq \pi_{i\infty}(\pi_\infty(m_2))$, and in turn $\pi_\infty(m_1) \neq \pi_\infty(m_2)$.

We remark that, in a sense, the second condition in the previous proposition just says that the covering $\mathcal{U}_i$ contains 'enough small open sets', a condition one would expect in the process of recovering M by a refinement of the coverings.

As alluded to before, there is a nice characterization of the points of M (or better still of $\pi_\infty(M)$) as the set of all closed points of P_∞. We have indeed a further Proposition; for the easy but long proof one is referred to [121],

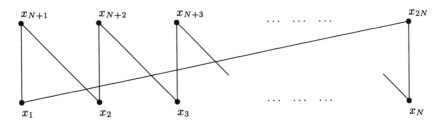

Fig. 3.8. The Hasse diagram for $P_{2N}(S^1)$

Proposition 3.3.3. *Let M be T_1 and let the collection $\{\mathcal{U}_i\}$ of coverings fulfill the 'fineness' condition of Proposition 3.3.2. Let each covering $\mathcal{U}_i$ consist only of sets which are bounded (have compact closure). Then $\pi_\infty : M \to P_\infty$ embeds M in P_∞ as the subspace of closed points.*

We remark that the additional requirement on the element of each covering is automatically fulfilled if M is compact.

As for the extra points of P_∞, one can prove that for any extra $y \in P_\infty$, there exists an $x \in \pi_\infty(M)$ to which y is 'infinitely close'. Indeed, P_∞ can be turned into a poset by defining a partial order relation as follows

$$x \preceq_\infty y \quad \Leftrightarrow \quad x_i \preceq y_i , \quad \forall\, i , \qquad (3.33)$$

where the coherent sequences $x = (x_i)$ and $y = (y_i)$ are any two elements of P_∞.[6] Then one can characterize $\pi_\infty(M)$ as the *set of maximal elements of P_∞*, with respect to the order $\preceq_\infty$. Given any such maximal element x, the points of P_∞ which are infinitely close to x are all (non maximal) points which converge to x, namely all (not maximal) $y \in P_\infty$ such that $y \preceq_\infty x$. In P_∞, these points y cannot be separated from the corresponding x. By identifying points in P_∞ which cannot be separated one recovers M. The interpretation that emerges is that the top points of a poset $P(M)$ (which are always closed) approximate the points of M and give all of M in the limit. The rôle of the remaining points is to 'glue' the top points together so as to produce a topologically nontrivial approximation to M. They also give the extra points in the limit. Figure 3.8 shows the $2N$-poset approximation to S^1 obtained with a covering consisting of N open sets. In Fig. 3.9 we have the associated projective system of posets. As seen in that Figure, by going from one level to the next one, only one of the bottom points x is 'split' in three $\{x_0, x_1, x_1\}$ while the others are not changed. The projection from one level to the previous one is the map which sends the triple $\{x_0, x_1, x_1\}$ to the parent x while acting as the identity on the remaining points. The projection

---

[6] In fact, one could directly construct P_∞ as the projective limit of a projective system of posets by defining a partial order on the coherent sequences as in (3.33).

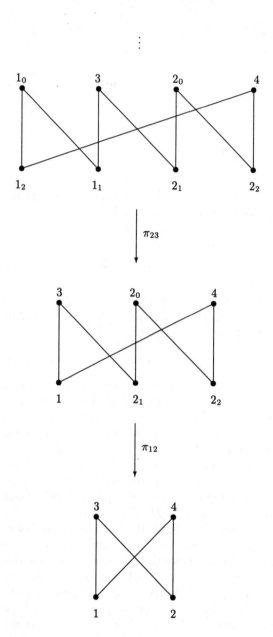

Fig. 3.9. The projective system of posets for S^1

is easily seen to be order preserving (and then continuous). As in the general case, the limit space P_∞ consists of S^1 together with extra points. These extra points come in couples anyone of which is 'glued' (in the sense of being infinitely close) to a point in a numerable collection of points. This collection is dense in S^1 and could be taken as the collection of all points of the form $\{m/2^n,\ m, n \in \mathbb{N}\}$ of the interval $[0, 1]$ with endpoints identified.

In [11] a somewhat different interpretation of the approximation and of the limiting procedure in terms of simplicial decompositions has been proposed.

3.4 Noncommutative Lattices

It turns out that any (finite) poset P is the structure space $\widehat{A}$ (the space of irreducible representations, see Sect. 2.3) of a noncommutative C^*-algebra A of operator valued functions which then plays the rôle of the algebra of continuous functions on P.[7] Indeed, there is a complete classification of all separable[8] C^*-algebras with a finite dual [8]. Given any finite T_0-space P, it is possible to construct a C^*-algebra $A(P, d)$ of operators on a separable[9] Hilbert space $\mathcal{H}(P, d)$ which satisfies $\widehat{A(P, d)} = P$. Here d is a function on P with values in $\mathbb{N} \cup \infty$ which is called a *defector*. Thus there is more than one algebra with the same structure space. We refer to [8, 61] for the actual construction of the algebras together with extensions to countable posets. Here, we shall instead describe a more general class of algebras, namely approximately finite dimensional ones, a subclass of which is associated with posets. As the name suggests, these algebras can be approximated by finite dimensional algebras, a fact which has been used in the construction of physical models on posets as we shall describe in Chap. 11. They are also useful in the analysis of the K-theory of posets as we shall see in Chap. 5.

Before we proceed, we mention that if a separable C^*-algebra has a finite dual than it is postliminal [8]. From Sect. 2.4 we know that for any such algebra A, irreducible representations are completely characterized by their kernels so that the structure space $\widehat{A}$ is homeomorphic with the space $Prim A$ of primitive ideals. As we shall see momentarily, the Jacobson topology on $Prim A$ is equivalent to the partial order defined by the inclusion of ideals. This fact in a sense 'closes the circle' making any poset, when thought of

---

[7] It is worth noticing that, a poset P being non Hausdorff, there cannot be 'enough' $\mathbb{C}$-valued continuous functions on P since the latter separate points. For instance, on the poset of Fig. 3.1 or Fig. 3.3 the only $\mathbb{C}$-valued continuous functions are the constant ones. In fact, the previous statement is true for each connected component of any poset.

[8] We recall that a C^*-algebra A is called *separable* if it admits a countable subset which is dense in the norm topology of A.

[9] Much as in the previous footnote, a Hilbert space $\mathcal{H}$ is called *separable* if it admits a countable basis.

as the $Prim\mathcal{A}$ space of a noncommutative algebra, a truly noncommutative space or, rather, a *noncommutative lattice*.

3.4.1 The Space $Prim\mathcal{A}$ as a Poset

We recall that in Sect. 2.3.1 we introduced the natural T_0-topology (the Jacobson topology) on the space $Prim\mathcal{A}$ of primitive ideals of a noncommutative C^*-algebra $\mathcal{A}$. In particular, from Prop. 2.3.4, we have that, given any subset W of $Prim\mathcal{A}$,

$$W \text{ is closed} \quad \Leftrightarrow \quad \mathcal{I} \in W \text{ and } \mathcal{I} \subseteq \mathcal{J} \Rightarrow \mathcal{J} \in W \ . \qquad (3.34)$$

Now, a partial order $\preceq$ is naturally introduced on $Prim\mathcal{A}$ by inclusion,

$$\mathcal{I}_1 \preceq \mathcal{I}_2 \quad \Leftrightarrow \quad \mathcal{I}_1 \subseteq \mathcal{I}_2 \ , \quad \forall\, \mathcal{I}_1, \mathcal{I}_2 \in Prim\mathcal{A} \ . \qquad (3.35)$$

From what we said after (3.14), given any subset W of the topological space $(Prim\mathcal{A}, \preceq)$,

$$W \text{ is closed} \quad \Leftrightarrow \quad \mathcal{I} \in W \text{ and } \mathcal{I} \preceq \mathcal{J} \Rightarrow \mathcal{J} \in W \ , \qquad (3.36)$$

which is just the partial order reading of (3.34). We infer that on $Prim\mathcal{A}$ the Jacobson topology and the partial order topology can be identified.

3.4.2 AF-Algebras

In this Section we shall describe approximately finite dimensional algebras following [16]. A general algebra of this sort may have a rather complicated ideal structure and a complicated primitive ideal structure. As mentioned before, for applications to posets only a special subclass is selected.

Definition 3.4.1. *A C^*-algebra $\mathcal{A}$ is said to be approximately finite dimensional (AF) if there exists an increasing sequence*

$$\mathcal{A}_0 \overset{I_0}{\hookrightarrow} \mathcal{A}_1 \overset{I_1}{\hookrightarrow} \mathcal{A}_2 \overset{I_2}{\hookrightarrow} \cdots \overset{I_{n-1}}{\hookrightarrow} \mathcal{A}_n \overset{I_n}{\hookrightarrow} \cdots \qquad (3.37)$$

of finite dimensional C^*-subalgebras of $\mathcal{A}$, such that $\mathcal{A}$ is the norm closure of $\bigcup_n \mathcal{A}_n$, $\mathcal{A} = \overline{\bigcup_n \mathcal{A}_n}$. The maps I_n are injective $*$-morphisms.

The algebra $\mathcal{A}$ is the *inductive* (or *direct*) *limit* of the *inductive* (or *direct*) *system* $\{\mathcal{A}_n, I_n\}_{n \in \mathbb{N}}$ of algebras [108, 131]. As a set, $\bigcup_n \mathcal{A}_n$ is made of coherent sequences,

$$\bigcup_n \mathcal{A}_n = \{a = (a_n)_{n \in \mathbb{N}} \ , a_n \in \mathcal{A}_n \mid \exists N_0 \ , a_{n+1} = I_n(a_n) \ , \forall\, n > N_0\}. \qquad (3.38)$$

Now the sequence $(\|a_n\|_{\mathcal{A}_n})_{n \in \mathbb{N}}$ is eventually decreasing since $\|a_{n+1}\| \leq \|a_n\|$ (the maps I_n are norm decreasing) and therefore convergent. One writes for the norm on $\mathcal{A}$,

$$\|(a_n)_{n\in\mathbb{N}}\| = \lim_{n\to\infty} \|a_n\|_{\mathcal{A}_n} . \tag{3.39}$$

Since the maps I_n are injective, the expression (3.39) gives a true norm directly and not simply a seminorm and there is no need to quotient out the zero norm elements.

We shall assume that the algebra $\mathcal{A}$ has a unit $\mathbb{I}$. If $\mathcal{A}$ and $\mathcal{A}_n$ are as before, then $\mathcal{A}_n + \mathbb{C}\mathbb{I}$ is clearly a finite dimensional C^*-subalgebra of $\mathcal{A}$ and moreover, $\mathcal{A}_n \subset \mathcal{A}_n + \mathbb{C}\mathbb{I} \subset \mathcal{A}_{n+1} + \mathbb{C}\mathbb{I}$. We may thus assume that each $\mathcal{A}_n$ contains the unit $\mathbb{I}$ of $\mathcal{A}$ and that the maps I_n are unital.

Example 3.4.1. Let $\mathcal{H}$ be an infinite dimensional (separable) Hilbert space. The algebra

$$\mathcal{A} = \mathcal{K}(\mathcal{H}) + \mathbb{C}\mathbb{I}_{\mathcal{H}} , \tag{3.40}$$

with $\mathcal{K}(\mathcal{H})$ the algebra of compact operators, is an AF-algebra [16]. The approximating algebras are given by

$$\mathcal{A}_n = \mathbb{M}_n(\mathbb{C}) \oplus \mathbb{C} , \quad n > 0 , \tag{3.41}$$

with embedding

$$\mathbb{M}_n(\mathbb{C}) \oplus \mathbb{C} \ni (\Lambda, \lambda) \mapsto \left(\left\{ \begin{matrix} \Lambda & 0 \\ 0 & \lambda \end{matrix} \right\}, \lambda \right) \in \mathbb{M}_{n+1}(\mathbb{C}) \oplus \mathbb{C} . \tag{3.42}$$

Indeed, let $\{\xi_n\}_{n\in\mathbb{N}}$ be an orthonormal basis in $\mathcal{H}$ and let $\mathcal{H}_n$ be the subspace generated by the first n basis elements, $\{\xi_1, \cdots, \xi_n\}$. With $\mathcal{P}_n$ the orthogonal projection onto $\mathcal{H}_n$, define

$$\begin{aligned} \mathcal{A}_n &= \{T \in \mathcal{B}(\mathcal{H}) \mid T(\mathbb{I} - \mathcal{P}_n) = (\mathbb{I} - \mathcal{P}_n)T \in \mathbb{C}(\mathbb{I} - \mathcal{P}_n)\} \\ &\simeq \mathcal{B}(\mathcal{H}_n) \oplus \mathbb{C} \simeq \mathbb{M}_n(\mathbb{C}) \oplus \mathbb{C} . \end{aligned} \tag{3.43}$$

Then $\mathcal{A}_n$ embeds in $\mathcal{A}_{n+1}$ as in (3.42). Since each $T \in \mathcal{A}_n$ is a sum of a finite rank operator and a multiple of the identity, one has that $\mathcal{A}_n \subseteq \mathcal{A} = \mathcal{K}(\mathcal{H}) + \mathbb{C}\mathbb{I}_{\mathcal{H}}$ and, in turn, $\overline{\bigcup_n \mathcal{A}_n} \subseteq \mathcal{A} = \mathcal{K}(\mathcal{H}) + \mathbb{C}\mathbb{I}_{\mathcal{H}}$. Conversely, since finite rank operators are norm dense in $\mathcal{K}(\mathcal{H})$, and finite linear combinations of strings $\{\xi_1, \cdots, \xi_n\}$ are dense in $\mathcal{H}$, one gets that $\mathcal{K}(\mathcal{H}) + \mathbb{C}\mathbb{I}_{\mathcal{H}} \subset \overline{\bigcup_n \mathcal{A}_n}$.

The algebra (3.40) has only two irreducible representations [8],

$$\begin{aligned} \pi_1 &: \mathcal{A} \longrightarrow \mathcal{B}(\mathcal{H}) , \quad a = (k + \lambda\mathbb{I}_{\mathcal{H}}) \mapsto \pi_1(a) = a , \\ \pi_2 &: \mathcal{A} \longrightarrow \mathcal{B}(\mathbb{C}) \simeq \mathbb{C} , \quad a = (k + \lambda\mathbb{I}_{\mathcal{H}}) \mapsto \pi_2(a) = \lambda , \end{aligned} \tag{3.44}$$

with $\lambda_1, \lambda_2 \in \mathbb{C}$ and $k \in \mathcal{K}(\mathcal{H})$; the corresponding kernels being

$$\begin{aligned} \mathcal{I}_1 &=: ker(\pi_1) = \{0\} , \\ \mathcal{I}_2 &=: ker(\pi_2) = \mathcal{K}(\mathcal{H}) . \end{aligned} \tag{3.45}$$

The partial order given by the inclusions $\mathcal{I}_1 \subset \mathcal{I}_2$ produces the two point poset shown in Fig. 3.10. As we shall see, this space is really the fundamental building block for all posets. A comparison with the poset of the line in Fig. 3.2, shows that it can be thought of as a two point approximation of an interval.

Fig. 3.10. The two point poset of the interval

In general, each subalgebra $\mathcal{A}_n$ being a finite dimensional C^*-algebra, is a direct sum of matrix algebras,

$$\mathcal{A}_n = \bigoplus_{k=1}^{k_n} \mathrm{M}_{d_k^{(n)}}(\mathbb{C}) \ , \tag{3.46}$$

where $\mathrm{M}_d(\mathbb{C})$ is the algebra of $d \times d$ matrices with complex coefficients. In order to study the embedding $\mathcal{A}_1 \hookrightarrow \mathcal{A}_2$ of any two such algebras $\mathcal{A}_1 = \bigoplus_{j=1}^{n_1} \mathrm{M}_{d_j^{(1)}}(\mathbb{C})$ and $\mathcal{A}_2 = \bigoplus_{k=1}^{n_2} \mathrm{M}_{d_k^{(2)}}(\mathbb{C})$, the following proposition [56, 131] is useful.

Proposition 3.4.1. *Let A and B be the direct sum of two matrix algebras,*

$$A = \mathrm{M}_{p_1}(\mathbb{C}) \oplus \mathrm{M}_{p_2}(\mathbb{C}) \ , \quad B = \mathrm{M}_{q_1}(\mathbb{C}) \oplus \mathrm{M}_{q_2}(\mathbb{C}) \ . \tag{3.47}$$

Then, any (unital) morphism $\alpha : A \to B$ can be written as the direct sum of the representations $\alpha_j : A \to \mathrm{M}_{q_j}(\mathbb{C}) \simeq \mathcal{B}(\mathbb{C}^{q_j}), j = 1, 2$. If π_{ji} is the unique irreducible representation of $\mathrm{M}_{p_i}(\mathbb{C})$ in $\mathcal{B}(\mathbb{C}^{q_j})$, then α_j splits into a direct sum of the π_{ji}'s with multiplicity N_{ji}, the latter being nonnegative integers.

Proof. This proposition just says that, by suppressing the symbols π_{ji}, and modulo a change of basis, the morphism $\alpha : A \to B$ is of the form

$$A \bigoplus B \mapsto \underbrace{A \oplus \cdots \oplus A}_{N_{11}} \oplus \underbrace{B \oplus \cdots \oplus B}_{N_{12}} \bigoplus \underbrace{A \oplus \cdots \oplus A}_{N_{21}} \oplus \underbrace{B \oplus \cdots \oplus B}_{N_{22}} \ ,$$
$$\tag{3.48}$$

with $A \bigoplus B \in \mathcal{A}$. Moreover, the dimensions (p_1, p_2) and (q_1, q_2) are related by

$$\begin{aligned} N_{11}p_1 + N_{12}p_2 &= q_1 \ , \\ N_{21}p_1 + N_{22}p_2 &= q_2 \ . \end{aligned} \tag{3.49}$$

Given a unital embedding $\mathcal{A}_1 \hookrightarrow \mathcal{A}_2$ of the algebras $\mathcal{A}_1 = \bigoplus_{j=1}^{n_1} \mathrm{M}_{d_j^{(1)}}(\mathbb{C})$ and $\mathcal{A}_2 = \bigoplus_{k=1}^{n_2} \mathrm{M}_{d_k^{(2)}}(\mathbb{C})$, by making use of Proposition 3.4.1 one can always choose suitable bases in $\mathcal{A}_1$ and $\mathcal{A}_2$ in such a way as to identify $\mathcal{A}_1$ with a subalgebra of $\mathcal{A}_2$ having the following form

$$\mathcal{A}_1 \simeq \bigoplus_{k=1}^{n_2} \left(\bigoplus_{j=1}^{n_1} N_{kj} \mathbb{M}_{d_j^{(1)}}(\mathbb{C}) \right) . \tag{3.50}$$

Here, with any two nonnegative integers p, q, the symbol $p\mathbb{M}_q(\mathbb{C})$ stands for

$$p\mathbb{M}_q(\mathbb{C}) \simeq \mathbb{M}_q(\mathbb{C}) \otimes_{\mathbb{C}} \mathbb{I}_p , \tag{3.51}$$

and one identifies $\bigoplus_{j=1}^{n_1} N_{kj} \mathbb{M}_{d_j^{(1)}}(\mathbb{C})$ with a subalgebra of $\mathbb{M}_{d_k^{(2)}}(\mathbb{C})$. The nonnegative integers N_{kj} satisfy the condition

$$\sum_{j=1}^{n_1} N_{kj} d_j^{(1)} = d_k^{(2)} . \tag{3.52}$$

One says that the algebra $\mathbb{M}_{d_j^{(1)}}(\mathbb{C})$ is *partially embedded* in $\mathbb{M}_{d_k^{(2)}}(\mathbb{C})$ with *multiplicity* N_{kj}. A useful way of representing the algebras $\mathcal{A}_1$, $\mathcal{A}_2$ and the embedding $\mathcal{A}_1 \hookrightarrow \mathcal{A}_2$ is by means of a diagram, the so called *Bratteli diagram* [16], which can be constructed out of the dimensions $d_j^{(1)}$, $j = 1, \ldots, n_1$ and $d_k^{(2)}$, $k = 1, \ldots, n_2$, of the diagonal blocks of the two algebras and out of the numbers N_{kj} that describe the partial embeddings. One draws two horizontal rows of vertices, the top (bottom) one representing $\mathcal{A}_1$ ($\mathcal{A}_2$) and consisting of n_1 (n_2) vertices, one for each block which are labeled by the corresponding dimensions $d_1^{(1)}, \ldots, d_{n_1}^{(1)}$ ($d_1^{(2)}, \ldots, d_{n_2}^{(2)}$). Then, for each $j = 1, \ldots, n_1$ and $k = 1, \ldots, n_2$, one has a relation $d_j^{(1)} \searrow^{N_{kj}} d_k^{(2)}$ to denote the fact that $\mathbb{M}_{d_j^{(1)}}(\mathbb{C})$ is embedded in $\mathbb{M}_{d_k^{(2)}}(\mathbb{C})$ with multiplicity N_{kj}.

For any AF-algebra $\mathcal{A}$ one repeats the procedure for each level, and in this way one obtains a semi-infinite diagram, denoted by $\mathcal{D}(\mathcal{A})$ which completely defines $\mathcal{A}$ up to isomorphism. The diagram $\mathcal{D}(\mathcal{A})$ depends not only on the collection of $\mathcal{A}$'s but also on the particular sequence $\{\mathcal{A}_n\}_{n\in\mathbb{N}}$ which generates $\mathcal{A}$. However, one can obtain an algorithm which allows one to construct from a given diagram all diagrams which define AF-algebras which are isomorphic with the original one [16]. The problem of identifying the limit algebra or of determining whether or not two such limits are isomorphic can be very subtle. Elliot [58] has devised an invariant for AF-algebras in terms of the corresponding K theory which completely distinguishes among them (see also [56]). We shall elaborate a bit on this in Chap. 5. It is worth remarking that the isomorphism class of an AF-algebra $\bigcup_n \mathcal{A}_n$ depends not only on the collection of algebras $\mathcal{A}_n$'s but also on the way they are embedded into each other.

Any AF-algebra is clearly separable but the converse is not true. Indeed, one can prove that a separable C^*-algebra $\mathcal{A}$ is an AF-algebra if and only if and it has the following approximation property: for each finite set $\{a_1, \ldots, a_n\}$ of elements of $\mathcal{A}$ and $\varepsilon > 0$, there exists a finite dimensional C^*-algebra $\mathcal{B} \subseteq \mathcal{A}$ and elements $b_1, \ldots, b_n \in \mathcal{B}$ such that $\|a_k - b_k\| < \varepsilon$, $k = 1, \ldots, n$.

Given a set $\mathcal{D}$ of ordered pairs $(n,k), k = 1, \cdots, k_n$, $n = 0, 1, \cdots$, with $k_0 = 1$, and a sequence $\{\searrow^p\}_{p=0,1,\cdots}$ of relations on $\mathcal{D}$, the latter is the diagram $\mathcal{D}(\mathcal{A})$ of an AF-algebras when the following conditions are satisfied,

(i) If $(n,k), (m,q) \in \mathcal{D}$ and $m = n + 1$, there exists one and only one nonnegative (or equivalently, at most a positive) integer p such that $(n,k) \searrow^p (n+1, q)$.
(ii) If $m \neq n + 1$, no such integer exists.
(iii) If $(n,k) \in \mathcal{D}$, there exists $q \in \{1, \cdots, n_{n+1}\}$ and a nonnegative integer p such that $(n,k) \searrow^p (n+1, q)$.
(iv) If $(n,k) \in \mathcal{D}$ and $n > 0$, there exists $q \in \{1, \cdots, n_{n-1}\}$ and a nonnegative integer p such that $(n-1, q) \searrow^p (n, k)$.

It is easy to see that the diagram of a given AF-algebra satisfies the previous conditions. Conversely, if the set $\mathcal{D}$ of ordered pairs satisfies these properties, one constructs by induction a sequence of finite dimensional C^*-algebras $\{\mathcal{A}_n\}_{n \in \mathbb{N}}$ and of injective morphisms $I_n : \mathcal{A}_n \to \mathcal{A}_{n+1}$ in such a manner so that the inductive limit $\{\mathcal{A}_n, I_n\}_{n \in \mathbb{N}}$ will have $\mathcal{D}$ as its diagram. Explicitly, one defines

$$\mathcal{A}_n = \bigoplus_{k;(n,k)\in\mathcal{D}} \mathrm{M}_{d_k^{(n)}}(\mathbb{C}) = \bigoplus_{k=1}^{k_n} \mathrm{M}_{d_k^{(n)}}(\mathbb{C}) , \qquad (3.53)$$

and morphisms

$$I_n : \bigoplus_{j=1}^{j_n} \mathrm{M}_{d_j^{(n)}}(\mathbb{C}) \longrightarrow \bigoplus_{k=1}^{k_{n+1}} \mathrm{M}_{d_k^{(n+1)}}(\mathbb{C}) ,$$
$$A_1 \oplus \cdots \oplus A_{j_n} \mapsto (\oplus_{j=1}^{j_n} N_{1j} A_j) \oplus \cdots \oplus (\oplus_{j=1}^{j_n} N_{k_{n+1}j} A_j) ,$$
$$(3.54)$$

where the integers N_{kj} are such that $(n,j) \searrow^{N_{kj}} (n+1, k)$ and we have used the notation (3.51). Notice that the dimension $d_k^{(n+1)}$ of the factor $\mathrm{M}_{d_k^{(n+1)}}(\mathbb{C})$ is not arbitrary but it is determined by a relation like (3.52),

$$d_k^{(n+1)} = \sum_{j=1}^{j_n} N_{kj} d_j^{(n)} . \qquad (3.55)$$

Example 3.4.2. An AF-algebra $\mathcal{A}$ is commutative if and only if all the factors $\mathrm{M}_{d_k^{(n)}}(\mathbb{C})$ are one dimensional, $\mathrm{M}_{d_k^{(n)}}(\mathbb{C}) \simeq \mathbb{C}$. Thus the corresponding diagram $\mathcal{D}$ has the property that for each $(n,k) \in \mathcal{D}, n > 0$, there is exactly one $(n-1, j) \in \mathcal{D}$ such that $(n-1, j) \searrow^1 (n, k)$.

There is a very nice characterization of commutative AF-algebras and of their primitive spectra [17],

Proposition 3.4.2. *Let $\mathcal{A}$ be a commutative C^*-algebra with unit $\mathbb{I}$. Then the following statements are equivalent.*

(i) The algebra $\mathcal{A}$ is AF.

(ii) The algebra $\mathcal{A}$ is generated in the norm topology by a sequence of projectors $\{\mathcal{P}_i\}$, with $\mathcal{P}_0 = \mathbb{I}$.

(iii) The space $Prim\mathcal{A}$ is a second-countable, totally disconnected, compact Hausdorff space.[10]

Proof. The equivalence of (i) and (ii) is clear. To prove that (iii) implies (ii), let X be a second-countable, totally disconnected, compact Hausdorff space. Then X has a countable basis $\{X_n\}$ of open-closed sets. Let $\mathcal{P}_n$ be the characteristic function of X_n. The *-algebra generated by the projector $\{\mathcal{P}_n\}$ is dense in $C(X)$: since $Prim C(X) = X$, (iii) implies (ii). The converse, that (ii) implies (iii), follows from the fact that projectors in a commutative C^*-algebra $\mathcal{A}$ correspond to open-closed subsets in its primitive space $Prim\mathcal{A}$.

Example 3.4.3. Let us consider the subalgebra $\mathcal{A}$ of the algebra $\mathcal{B}(\mathcal{H})$ of bounded operators on an infinite dimensional (separable) Hilbert space $\mathcal{H} = \mathcal{H}_1 \oplus \mathcal{H}_2$, given in the following manner. Let $\mathcal{P}_j$ be the projection operators on $\mathcal{H}_j$, $j = 1, 2$, and $\mathcal{K}(\mathcal{H})$ the algebra of compact operators on $\mathcal{H}$. Then, the algebra $\mathcal{A}$ is

$$\mathcal{A}_\vee = \mathbb{C}\mathcal{P}_1 + \mathcal{K}(\mathcal{H}) + \mathbb{C}\mathcal{P}_2 . \tag{3.56}$$

The use of the symbol $\mathcal{A}_\vee$ is due to the fact that, as we shall see below, this algebra is associated with any part of the poset of the line in Fig. 3.2, of the form

$$\bigvee = \{y_{i-1}, x_i, y_i\} , \tag{3.57}$$

in the sense that this poset is identified with the space of primitive ideals of $\mathcal{A}_\vee$. The C^*-algebra (3.56) can be obtained as the inductive limit of the following sequence of finite dimensional algebras:

$$\begin{aligned}
\mathcal{A}_0 &= \mathbb{M}_1(\mathbb{C}) , \\
\mathcal{A}_1 &= \mathbb{M}_1(\mathbb{C}) \oplus \mathbb{M}_1(\mathbb{C}) , \\
\mathcal{A}_2 &= \mathbb{M}_1(\mathbb{C}) \oplus \mathbb{M}_2(\mathbb{C}) \oplus \mathbb{M}_1(\mathbb{C}) , \\
\mathcal{A}_3 &= \mathbb{M}_1(\mathbb{C}) \oplus \mathbb{M}_4(\mathbb{C}) \oplus \mathbb{M}_1(\mathbb{C}) , \\
&\vdots \\
\mathcal{A}_n &= \mathbb{M}_1(\mathbb{C}) \oplus \mathbb{M}_{2n-2}(\mathbb{C}) \oplus \mathbb{M}_1(\mathbb{C}) , \\
&\vdots
\end{aligned} \tag{3.58}$$

where, for $n \geq 1$, $\mathcal{A}_n$ is embedded in $\mathcal{A}_{n+1}$ as follows

---

[10] We recall that a topological space is called *totally disconnected* if the connected component of each point consists only of the point itself. Also, a topological space is called *second-countable* if it admits a countable basis of open sets.

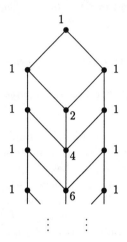

Fig. 3.11. The Bratteli diagram of the algebra $\mathcal{A}_\mathsf{V}$; the labels indicate the dimension of the corresponding matrix algebras

$$M_1(\mathbb{C}) \oplus M_{2n-2}(\mathbb{C}) \oplus M_1(\mathbb{C}) \hookrightarrow$$
$$\hookrightarrow M_1(\mathbb{C}) \oplus (M_1(\mathbb{C}) \oplus M_{2n-2}(\mathbb{C}) \oplus M_1(\mathbb{C})) \oplus M_1(\mathbb{C}) ,$$

$$
\begin{bmatrix} \lambda_1 & 0 & 0 \\ 0 & B & 0 \\ 0 & 0 & \lambda_2 \end{bmatrix} \mapsto
\begin{bmatrix} \lambda_1 & 0 & 0 & 0 & 0 \\ 0 & \lambda_1 & 0 & 0 & 0 \\ 0 & 0 & B & 0 & 0 \\ 0 & 0 & 0 & \lambda_2 & 0 \\ 0 & 0 & 0 & 0 & \lambda_2 \end{bmatrix} , \tag{3.59}
$$

for any $\lambda_1, \lambda_2 \in M_1(\mathbb{C})$ and any $B \in M_{2n-2}(\mathbb{C})$. The corresponding Bratteli diagram is shown in Fig. 3.11. The algebra (3.56) has three irreducible representations,

$$
\begin{aligned}
\pi_1 &: \mathcal{A}_\mathsf{V} \longrightarrow \mathcal{B}(\mathcal{H}) , & a &= (\lambda_1 \mathcal{P}_1 + k + \lambda_2 \mathcal{P}_2) \mapsto \pi_1(a) = a , \\
\pi_2 &: \mathcal{A}_\mathsf{V} \longrightarrow \mathcal{B}(\mathbb{C}) \simeq \mathbb{C} , & a &= (\lambda_1 \mathcal{P}_1 + k + \lambda_2 \mathcal{P}_2) \mapsto \pi_2(a) = \lambda_1 , \\
\pi_3 &: \mathcal{A}_\mathsf{V} \longrightarrow \mathcal{B}(\mathbb{C}) \simeq \mathbb{C} , & a &= (\lambda_1 \mathcal{P}_1 + k + \lambda_2 \mathcal{P}_2) \mapsto \pi_3(a) = \lambda_2 ,
\end{aligned}
\tag{3.60}
$$

with $\lambda_1, \lambda_2 \in \mathbb{C}$ and $k \in \mathcal{K}(\mathcal{H})$. The corresponding kernels are

$$
\begin{aligned}
\mathcal{I}_1 &= \{0\} , \\
\mathcal{I}_2 &= \mathcal{K}(\mathcal{H}) + \mathbb{C}\mathcal{P}_2 , \\
\mathcal{I}_3 &= \mathbb{C}\mathcal{P}_1 + \mathcal{K}(\mathcal{H}) .
\end{aligned}
\tag{3.61}
$$

The partial order given by the inclusions $\mathcal{I}_1 \subset \mathcal{I}_2$ and $\mathcal{I}_1 \subset \mathcal{I}_3$ (which, as shown in Sect. 3.4.1 is an equivalent way to provide the Jacobson topology) produces a topological space $Prim\mathcal{A}_\mathsf{V}$ which is just the V poset in (3.57).

3.4.3 From Bratteli Diagrams to Noncommutative Lattices

From the Bratteli diagram of an AF-algebra $\mathcal{A}$ one can also obtain the (norm closed two-sided) ideals of the latter and determine which ones are primitive. On the set of such ideals the topology is then given by constructing a poset whose partial order is provided by the inclusion of ideals. Therefore, both $Prim(\mathcal{A})$ and its topology can be determined from the Bratteli diagram of $\mathcal{A}$. This is possible thanks to the following results of Bratteli [16].

Proposition 3.4.3. *Let $\mathcal{A} = \overline{\bigcup_n \mathcal{U}_n}$ be any AF-algebra with associated Bratteli diagram $\mathcal{D}(\mathcal{A})$. Let $\mathcal{I}$ be an ideal of $\mathcal{A}$. Then $\mathcal{I}$ has the form*

$$\mathcal{I} = \overline{\bigcup_{n=1}^{\infty} \oplus_{k;(n,k)\in\Lambda_{\mathcal{I}}} \mathbb{M}_{d_k^{(n)}}(\mathbb{C})} \tag{3.62}$$

with the subset $\Lambda_{\mathcal{I}} \subset \mathcal{D}(\mathcal{A})$ satisfying the following two properties:

i) if $(n,k) \in \Lambda_{\mathcal{I}}$ and $(n,k) \searrow^p (n+1,j)$, $p > 0$, then $(n+1,j)$ belongs to $\Lambda_{\mathcal{I}}$;

ii) if all factors $(n+1,j)$, $j = 1,\ldots,n_{n+1}$, in which (n,k) is partially embedded belong to $\Lambda_{\mathcal{I}}$, then (n,k) belongs to $\Lambda_{\mathcal{I}}$.

Conversely, if $\Lambda \subset \mathcal{D}(\mathcal{A})$ satisfies properties (i) and (ii) above, then the subset $\mathcal{I}_\Lambda$ of $\mathcal{A}$ defined by (3.62) (with Λ substituted for $\Lambda_{\mathcal{I}}$) is an ideal in $\mathcal{A}$ such that $\mathcal{I} \cap \mathcal{A}_n = \oplus_{k;(n,k)\in\Lambda_{\mathcal{I}}} \mathbb{M}_{d_k^{(n)}}(\mathbb{C})$.

Proposition 3.4.4. *Let $\mathcal{A} = \overline{\bigcup_n \mathcal{U}_n}$, let $\mathcal{I}$ be an ideal of $\mathcal{A}$ and let $\Lambda_{\mathcal{I}} \subset \mathcal{D}(\mathcal{A})$ be the associated subdiagram. Then the following three conditions are equivalent.[11]*

(i) The ideal $\mathcal{I}$ is primitive.

(ii) There do not exist two ideals $\mathcal{I}_1, \mathcal{I}_2 \in \mathcal{A}$ such that $\mathcal{I}_1 \neq \mathcal{I} \neq \mathcal{I}_2$ and $\mathcal{I} = \mathcal{I}_1 \cap \mathcal{I}_2$.

(iii) If $(n,k), (m,q) \notin \Lambda_{\mathcal{I}}$, there exists an integer $p \geq n$, $p \geq m$, and a couple $(p,r) \notin \Lambda_{\mathcal{I}}$ such that $\mathbb{M}_{d_k^{(n)}}(\mathbb{C})$ and $\mathbb{M}_{d_q^{(m)}}(\mathbb{C})$ are both partially embedded in $\mathbb{M}_{d_r^{(p)}}(\mathbb{C})$ (equivalently, there are two sequences along the diagram $\mathcal{D}(\mathcal{A})$ starting at the points (n,k) and (m,q) with both ending at the same point (p,r)).

We recall that the whole of $\mathcal{A}$ is an ideal which, by definition, is not primitive since the trivial representation $\mathcal{A} \to 0$ is not irreducible. Furthermore, the ideal $\{0\} \subset \mathcal{A}$ is primitive if and only if $\mathcal{A}$ is primitive, which means that it has an irreducible faithful representation. This fact can also be inferred from the Bratteli diagram. Now, the ideal $\{0\}$, being represented at each level by the element $0 \in \mathcal{A}_n$[12], is not associated with any subdiagram of

---

[11] In fact, the equivalence of (i) and (ii) is true for any separable C^*-algebra [43].

[12] In fact one could think of $\Lambda_{\{0\}}$ as being the empty set.

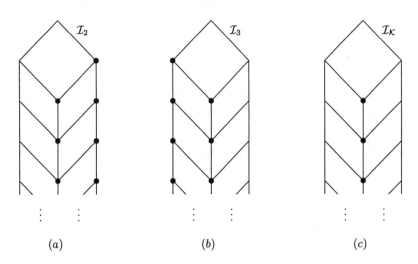

Fig. 3.12. The three ideals of the algebra $\mathcal{A}_\vee$

$\mathcal{D}(\mathcal{A})$. Therefore, to check if $\{0\}$ is primitive, we have the following corollary of Proposition 3.4.4.

Proposition 3.4.5. *Let $\mathcal{A} = \overline{\bigcup_n \mathcal{U}_n}$. Then the following conditions are equivalent.*

(i) The algebra $\mathcal{A}$ is primitive (the ideal $\{0\}$ is primitive).

(ii) There do not exist two ideals in $\mathcal{A}$ different from $\{0\}$ whose intersection is $\{0\}$.

(iii) If $(n, k), (m, q) \in \mathcal{D}(\mathcal{A})$, there exists an integer $p \geq n$, $p \geq m$, and a couple $(p, r) \in \mathcal{D}(\mathcal{A})$ such that $\mathbb{M}_{d_k^{(n)}}(\mathbb{C})$ and $\mathbb{M}_{d_q^{(m)}}(\mathbb{C})$ are both partially embedded in $\mathbb{M}_{d_r^{(p)}}(\mathbb{C})$ (equivalently, any two points of the diagram $\mathcal{D}(\mathcal{A})$ can be connected to a single point at a lower level of the diagram).

For instance, from the diagram of Fig. 3.11 we infer that the corresponding algebra is primitive, meaning that the ideal $\{0\}$ is primitive.

Example 3.4.4. As a simple example, consider the diagram of Fig. 3.11. The corresponding AF-algebra $\mathcal{A}_\vee$ in (3.56) contains only three nontrivial ideals whose diagrammatic representation is in Fig. 3.12.

In these pictures the points belonging to the same ideal are marked with a " $\bullet$ ". It is not difficult to check that only $\mathcal{I}_2$ and $\mathcal{I}_3$ are primitive ideals, since $\mathcal{I}_\mathcal{K}$ does not satisfy the property *(iii)* above. Now $\mathcal{I}_1 = \{0\}$ is an ideal which clearly belongs to both $\mathcal{I}_2$ and $\mathcal{I}_3$ so that $Prim(\mathcal{A})$ is any $\vee$ part of Fig. 3.2 of the form $\vee = \{y_{i-1}, x_i, y_i\}$. From the diagram of Fig. 3.12 one immediately obtains

$$\mathcal{I}_2 = \mathbb{C}\mathbb{I}_{\mathcal{H}} + \mathcal{K}(\mathcal{H}) \ ,$$
$$\mathcal{I}_1 = \mathbb{C}\mathbb{I}_{\mathcal{H}} + \mathcal{K}(\mathcal{H}) \ , \tag{3.63}$$

$\mathcal{H}$ being an infinite dimensional Hilbert space. Thus, $\mathcal{I}_2$ and $\mathcal{I}_3$ can be identified with the corresponding ideals of $\mathcal{A}_\vee$ given in (3.61). As for $\mathcal{I}_{\mathcal{K}}$, from Fig. 3.12 one gets $\mathcal{I}_{\mathcal{K}} = \mathcal{K}(\mathcal{H})$ which is not a primitive ideal of $\mathcal{A}_\vee$.

3.4.4 From Noncommutative Lattices to Bratteli Diagrams

There is also a reverse procedure which allows one to construct an AF-algebra (or rather its Bratteli diagram $\mathcal{D}(\mathcal{A})$) whose primitive ideal space is a given (finitary, noncommutative) lattice P [17, 18]. We shall briefly describe this procedure while referring to [60, 61] for more details and several examples.

Proposition 3.4.6. *Let P be a topological space with the following properties,*

(i) The space P is T_0;
(ii) If $F \subset P$ is a closed set which is not the union of two proper closed subsets, then F is the closure of a one-point set;
(iii) The space P contains at most a countable number of closed sets;
(iv) If $\{F_n\}_n$ is a decreasing ($F_{n+1} \subseteq F_n$) sequence of closed subsets of P, then $\bigcap_n F_n$ is an element in $\{F_n\}_n$.

Then, there exists an AF algebra $\mathcal{A}$ whose primitive space $Prim\mathcal{A}$ is homeomorphic to P.

Proof. The proof consists in constructing explicitly the Bratteli diagram $\mathcal{D}(\mathcal{A})$ of the algebra $\mathcal{A}$. We shall sketch the main steps while referring to [17, 18] for more details.

- Let $\{K_0, K_1, K_2, \ldots\}$ be the collection of all closed sets in the lattice P, with $K_0 = P$.

- Consider the subcollection $\mathcal{K}_n = \{K_0, K_1, \ldots, K_n\}$ and let $\mathcal{K}'_n$ be the smallest collection of (closed) sets in P containing $\mathcal{K}_n$ which is closed under union and intersection.

- Consider the algebra of sets[13] generated by the collection $\mathcal{K}_n$. Then, the minimal sets $\mathcal{Y}_n = \{Y_n(1), Y_n(2), \ldots, Y_n(k_n)\}$ of this algebra form a partition of P.

---

[13] We recall that a non empty collection R of subsets of a set X is called an *algebra of sets* if R is closed under the operations of union, i.e. $E, F \in R \Rightarrow E \cup F \in R$, and of complement, i.e. $E \in R \Rightarrow E^c =: X \setminus E \in R$.

- Let $F_n(j)$ be the smallest set in the subcollection $\mathcal{K}'_n$ which contains $Y_n(j)$. Define $\mathcal{F}_n = \{F_n(1), F_n(2), \ldots, F_n(k_n)\}$.

- As a consequence of the assumptions in the Proposition one has that

$$Y_n(k) \subseteq F_n(k) \ , \tag{3.64}$$

$$\bigcup_k Y_n(k) = P \ , \tag{3.65}$$

$$\bigcup_k F_n(k) = P \ , \tag{3.66}$$

$$Y_n(k) = F_n(k) \setminus \bigcup_{p \neq k} \{F_n(p) \mid F_n(p) \subset F_n(k)\} \ , \tag{3.67}$$

$$F_n(k) = \bigcup_p \{F_{n+1}(p) \mid F_{n+1}(p) \subseteq F_n(k)\} \ , \tag{3.68}$$

If $F \subset P$ is closed , $\exists \, n \geq 0$, s.t.

$$F_n(k) = \bigcup_p \{F_n(p) \mid F_n(p) \subseteq F\} \ . \tag{3.69}$$

- The diagram $\mathcal{D}(\mathcal{A})$ is constructed as follows.
  - (1.) *The n-th level of $\mathcal{D}(\mathcal{A})$ has k_n points, one for each set $Y_n(k)$, where $k = 1, \cdots, k_n$.*
    Thus $\mathcal{D}(\mathcal{A})$ is the set of all ordered pairs (n, k),
    $k = 1, \ldots, k_n$, $n = 0, 1, \ldots$.
  - (2.) *The point corresponding to $Y_n(k)$ at level n of the diagram is linked to the point corresponding to $Y_{n+1}(j)$ at level $n+1$, if and only if $Y_n(k) \cap F_{n+1}(j) \neq \emptyset$. The multiplicity of the embedding is always 1.*
    Thus, the partial embeddings of the diagram are given by

$$(n, k) \quad \searrow^p \quad (n+1, j) \ , \quad \text{with}$$
$$p = 1 \ \text{if} \ Y_n(k) \cap F_{n+1}(j) \neq \emptyset \ , \tag{3.70}$$
$$p = 0 \ \text{otherwise} \ .$$

That the diagram $\mathcal{D}(\mathcal{A})$ is really the diagram of an AF algebra $\mathcal{A}$, namely that conditions $(i) - (iv)$ of page 40 are satisfied, follows from the conditions (3.64)-(3.69) above.

Before we proceed, recall from Proposition (2.3.3) that there is a bijective correspondence between ideals in a C^*-algebra and closed sets in $Prim \mathcal{A}$, the correspondence being given by (2.36). We shall construct a similar correspondence between closed subsets $F \subseteq P$ and the ideals $\mathcal{I}_F$ in the AF-algebra $\mathcal{A}$ with subdiagram $\Lambda_F \subseteq \mathcal{D}(\mathcal{A})$. Given then, a closed subset $F \subseteq P$, from (3.69), there exists an m such that $F \subseteq \mathcal{K}'_m$. Define

$$(\Lambda_F)_n = \{(n, k) \mid n \geq m \ , \ Y_n(k) \cap F = \emptyset\} \ . \tag{3.71}$$

By using (3.67) one proves that conditions (i) and (ii) of Proposition 3.4.3 are satisfied. As a consequence, if Λ_F is the smallest subdiagram corresponding to an ideal $\mathcal{I}_F$, namely the smallest subdiagram satisfying conditions (i) and (ii) of Proposition 3.4.3, which also contains $(\Lambda_F)_n$, one has that

$$(\Lambda_F)_n = \Lambda_F \bigcap \{(n,k) \mid (n,k) \in \mathcal{D}(\mathcal{A}),\ n \geq m\} , \qquad (3.72)$$

which, in turn, implies that the mapping $F \mapsto \Lambda_F \leftrightarrow \mathcal{I}_F$ is injective.

To show surjectivity, let $\mathcal{I}$ be an ideal in $\mathcal{A}$ with associated subdiagram $\Lambda_{\mathcal{I}}$. For $n = 0, 1, \ldots$, define

$$F_n = P \setminus \bigcup_k \{Y_n(k) \mid \exists (n-1,p) \in \Lambda_{\mathcal{I}} ,\ (n-1,p) \searrow^1 (n,k) \in \Lambda_{\mathcal{I}}\} . \quad (3.73)$$

Then $\{F_n\}_n$ is a decreasing sequence of closed sets in P. By assumption (iv), there exists an m such that $F_m = \bigcap_n F_n$. By defining $F = F_m$, one has $F_n = F$ for $n \geq m$ and

$$\Lambda_{\mathcal{I}} \bigcap \{(n,k) \mid n \geq m\} =: (\Lambda_F)_m . \qquad (3.74)$$

Thus, $\Lambda_{\mathcal{I}} = \Lambda_F$ and the mapping $F \mapsto \mathcal{I}_F$ is surjective.

Finally, from the definition it follows that

$$F_1 \subseteq F_2 \iff \mathcal{I}_{F_1} \supseteq \mathcal{I}_{F_2} . \qquad (3.75)$$

For any point $x \in P$, the closure $\overline{\{x\}}$ is not the union of two proper closed subsets. From (3.75), the corresponding ideal $\mathcal{I}_{\overline{\{x\}}}$ is not the intersection of two ideals different from itself, thus it is primitive (see Proposition 3.4.4). Conversely, if $\mathcal{I}_F$ is primitive, it is not the intersection of two ideals different from itself, thus from (3.75) F is not the union of two proper closed subsets, and from assumption (ii), it is the closure of a one-point set. We have then proven that the ideal $\mathcal{I}_F$ is primitive if and only if F is the closure of a one-point set.

By taking into account the bijection between closed sets of the space P and ideals of the algebra $\mathcal{A}$ and the corresponding bijection between closed sets of the space $Prim\mathcal{A}$ and ideals of the algebra $\mathcal{A}$, we see that the bijection between P and $Prim\mathcal{A}$ which associates the corresponding primitive ideal to any point of P, is a homeomorphism.

We know that different algebras could yield the same space of primitive ideals (see the notion of strong Morita equivalence in App. A.4). It may happen that by changing the order in which the closed sets of P are taken in the construction of the previous proposition, one produces different algebras, all having the same space of primitive ideals though, and so all producing spaces which are homeomorphic to the starting P (any two of these spaces being, a fortiori, homeomorphic).

Example 3.4.5. As a simple example, consider again the lattice,

$$\bigvee = \{y_{i-1}, x_i, y_i\} \equiv \{x_2, x_1, x_3\} \ . \tag{3.76}$$

This topological space contains four closed sets:

$$K_0 = \{x_2, x_1, x_3\} \ , K_1 = \{x_2\} \ , K_2 = \{x_3\} \ , K_3 = \{x_2, x_3\} = K_1 \cup K_2 \ . \tag{3.77}$$

Thus, with the notation of Proposition 3.4.6, it is not difficult to check that:

$$\begin{array}{ll}
\mathcal{K}_0 = \{K_0\} \ , & \mathcal{K}'_0 = \{K_0\} \ , \\
\mathcal{K}_1 = \{K_0, K_1\} \ , & \mathcal{K}'_1 = \{K_0, K_1\} \ , \\
\mathcal{K}_2 = \{K_0, K_1, K_2\} \ , & \mathcal{K}'_2 = \{K_0, K_1, K_2, K_3\} \ , \\
\mathcal{K}_3 = \{K_0, K_1, K_2, K_3\} \ , & \mathcal{K}'_3 = \{K_0, K_1, K_2, K_3\} \ ,
\end{array}$$

$$\vdots$$

$$Y_0(1) = \{x_1, x_2, x_3\} \ , \qquad\qquad F_0(1) = K_0 \ ,$$

$$Y_1(1) = \{x_2\} \ , \quad Y_1(2) = \{x_1, x_3\} \ , \qquad F_1(1) = K_1 \ , \quad F_1(2) = K_0 \ ,$$

$$Y_2(1) = \{x_2\} \ , \quad Y_2(2) = \{x_1\} \ , \qquad F_2(1) = K_1 \ , \quad F_2(2) = K_0 \ ,$$
$$Y_2(3) = \{x_3\} \ , \qquad\qquad\qquad\qquad F_2(3) = K_2 \ ,$$

$$Y_3(1) = \{x_2\} \ , \quad Y_3(2) = \{x_1\} \ , \qquad F_3(1) = K_1 \ , \quad F_3(2) = K_0 \ ,$$
$$Y_3(3) = \{x_3\} \ , \qquad\qquad\qquad\qquad F_3(2) = K_2 \ ,$$

$$\vdots$$

$$\tag{3.78}$$

Since $\bigvee$ has only a finite number of points (three), and hence a finite number of closed sets (four), the partition of $\bigvee$ repeats itself after the third level. Figure 3.13 shows the corresponding diagram, obtained through rules (1.) and (2.) in Proposition 3.4.6 above (on page 46). By using the fact that the first matrix algebra $\mathcal{A}_0$ is $\mathbb{C}$ and the fact that all the embeddings have multiplicity one, the diagram of Fig. 3.13 is seen to coincide with the diagram of Fig. 3.11. As we have previously said, the latter corresponds to the AF-algebra

$$\mathcal{A}_{\bigvee} = \mathbb{C}P_1 + \mathcal{K}(\mathcal{H}) + \mathbb{C}P_2 \ , \quad \mathcal{H} = \mathcal{H}_1 \oplus \mathcal{H}_2 \ . \tag{3.79}$$

Example 3.4.6. Another interesting example is provided by the lattice $P_4(S^1)$ for the one-dimensional sphere in Fig. 3.1. This topological space contains six closed sets:

$$K_0 = \{x_1, x_2, x_3, x_4\} \ , \ K_1 = \{x_1, x_3, x_4\} \ , \ K_2 = \{x_3\} \ , \ K_3 = \{x_4\} \ ,$$
$$K_4 = \{x_2, x_3, x_4\} \ , \ K_5 = \{x_3, x_4\} = K_2 \cup K_3 \ . \tag{3.80}$$

Thus, with the notation of Proposition 3.4.6, one finds,

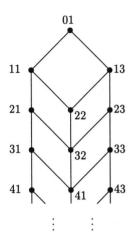

Fig. 3.13. The Bratteli diagram associated with the poset $\bigvee$; the label nk stands for $Y_n(k)$

$$
\begin{aligned}
\mathcal{K}_0 &= \{K_0\}\,, & \mathcal{K}'_0 &= \{K_0\}\,, \\
\mathcal{K}_1 &= \{K_0, K_1\}\,, & \mathcal{K}'_1 &= \{K_0, K_1\}\,, \\
\mathcal{K}_2 &= \{K_0, K_1, K_2\}\,, & \mathcal{K}'_2 &= \{K_0, K_1, K_2\}\,, \\
\mathcal{K}_3 &= \{K_0, K_1, K_2, K_3\}\,, & \mathcal{K}'_3 &= \{K_0, K_1, K_2, K_3, K_5\}\,, \\
\mathcal{K}_4 &= \{K_0, K_1, K_2, K_3, K_4\}\,, & \mathcal{K}'_4 &= \{K_0, K_1, K_2, K_3, K_4, K_5\}\,, \\
\mathcal{K}_5 &= \{K_0, K_1, K_2, K_3, K_4, K_5\}\,, & \mathcal{K}'_5 &= \{K_0, K_1, K_2, K_3, K_4, K_5\}\,, \\
\vdots &&&
\end{aligned}
$$

$$
\begin{aligned}
Y_0(1) &= \{x_1, x_2, x_3, x_4\}\,, & F_0(1) &= K_0\,, \\[4pt]
Y_1(1) &= \{x_1, x_3, x_4\}\,, && \\
& \quad Y_1(2) = \{x_2\}\,, & F_1(1) &= K_1\,, \quad F_1(2) = K_0\,, \\[4pt]
Y_2(1) &= \{x_3\}\,, \quad Y_2(2) = \{x_2\}\,, & F_2(1) &= K_2\,, \quad F_2(2) = K_0\,, \\
Y_2(3) &= \{x_1, x_4\}\,, & F_2(3) &= K_1\,, \\[4pt]
Y_3(1) &= \{x_3\}\,, \quad Y_3(2) = \{x_2\}\,, & F_3(1) &= K_2\,, \quad F_3(2) = K_0\,, \\
Y_3(3) &= \{x_1\}\,, \quad Y_3(4) = \{x_4\}\,, & F_3(3) &= K_1\,, \quad F_3(4) = K_3\,, \\[4pt]
Y_4(1) &= \{x_3\}\,, \quad Y_4(2) = \{x_2\}\,, & F_4(1) &= K_2\,, \quad F_4(2) = K_4\,, \\
Y_4(3) &= \{x_1\}\,, \quad Y_4(4) = \{x_4\}\,, & F_4(3) &= K_1\,, \quad F_4(4) = K_3\,, \\[4pt]
Y_5(1) &= \{x_3\}\,, \quad Y_5(2) = \{x_2\}\,, & F_5(1) &= K_2\,, \quad F_5(2) = K_4\,, \\
Y_5(3) &= \{x_1\}\,, \quad Y_5(4) = \{x_4\}\,, & F_5(3) &= K_1\,, \quad F_5(4) = K_3\,, \\
\vdots &&&
\end{aligned}
$$

$$(3.81)$$

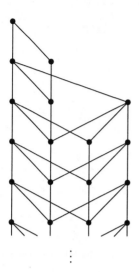

Fig. 3.14. The Bratteli diagram for the circle poset $P_4(S^1)$

Since there are a finite number of points (four), and hence a finite number of closed sets (six), the partition of $P_4(S^1)$ repeats itself after the fourth level. The corresponding Bratteli diagram is exhibited in Fig. 3.14. The ideal $\{0\}$ is not primitive. The algebra is given by

$$
\begin{aligned}
\mathcal{A}_0 &= \mathbb{M}_1(\mathbb{C}) , \\
\mathcal{A}_1 &= \mathbb{M}_1(\mathbb{C}) \oplus \mathbb{M}_1(\mathbb{C}) , \\
\mathcal{A}_2 &= \mathbb{M}_1(\mathbb{C}) \oplus \mathbb{M}_2(\mathbb{C}) \oplus \mathbb{M}_1(\mathbb{C}) , \\
\mathcal{A}_3 &= \mathbb{M}_1(\mathbb{C}) \oplus \mathbb{M}_4(\mathbb{C}) \oplus \mathbb{M}_2(\mathbb{C}) \oplus \mathbb{M}_1(\mathbb{C}) , \\
\mathcal{A}_4 &= \mathbb{M}_1(\mathbb{C}) \oplus \mathbb{M}_6(\mathbb{C}) \oplus \mathbb{M}_4(\mathbb{C}) \oplus \mathbb{M}_1(\mathbb{C}) ,
\end{aligned}
\tag{3.82}
$$

$$
\vdots
$$

$$
\mathcal{A}_n = \mathbb{M}_1(\mathbb{C}) \oplus \mathbb{M}_{2n-2}(\mathbb{C}) \oplus \mathbb{M}_{2n-4}(\mathbb{C}) \oplus \mathbb{M}_1(\mathbb{C}) ,
$$

$$
\vdots
$$

where, for $n > 2$, $\mathcal{A}_n$ is embedded in $\mathcal{A}_{n+1}$ as follows

$$
\begin{bmatrix} \lambda_1 & & \\ & B & \\ & & C \\ & & & \lambda_2 \end{bmatrix}
\mapsto
\begin{bmatrix}
\lambda_1 & & & & & & \\
& \begin{matrix} \lambda_1 & 0 & 0 \\ 0 & B & 0 \\ 0 & 0 & \lambda_2 \end{matrix} & & & \\
& & \begin{matrix} \lambda_1 & 0 & 0 \\ 0 & C & 0 \\ 0 & 0 & \lambda_2 \end{matrix} & \\
& & & & \lambda_2
\end{bmatrix} ,
\tag{3.83}
$$

with $\lambda_1, \lambda_2 \in \mathbb{M}_1(\mathbb{C})$, $B \in \mathbb{M}_{2n-2}(\mathbb{C})$ and $C \in \mathbb{M}_{2n-4}(\mathbb{C})$; elements which are not shown are equal to zero. The algebra limit $\mathcal{A}_{P_4(S^1)}$ can be realized explicitly as a subalgebra of bounded operators on an infinite dimensional Hilbert space $\mathcal{H}$ naturally associated with the poset $P_4(S^1)$. Firstly, to any *link* (x_i, x_j), $x_i \succ x_j$, of the poset one associates a Hilbert space $\mathcal{H}_{ij}$; for the case at hand, one has four Hilbert spaces, $\mathcal{H}_{31}, \mathcal{H}_{32}, \mathcal{H}_{41}, \mathcal{H}_{42}$. Then, since all links are at the same level, $\mathcal{H}$ is just given by the direct sum

$$\mathcal{H} = \mathcal{H}_{31} \oplus \mathcal{H}_{32} \oplus \mathcal{H}_{41} \oplus \mathcal{H}_{42} . \tag{3.84}$$

The algebra $\mathcal{A}_{P_4(S^1)}$ is given by [61],

$$\mathcal{A}_{P_4(S^1)} = \mathbb{C}\mathcal{P}_{\mathcal{H}_{31} \oplus \mathcal{H}_{32}} + \mathcal{K}_{\mathcal{H}_{31} \oplus \mathcal{H}_{41}} + \mathcal{K}_{\mathcal{H}_{32} \oplus \mathcal{H}_{42}} + \mathbb{C}\mathcal{P}_{\mathcal{H}_{41} \oplus \mathcal{H}_{42}} . \tag{3.85}$$

Here $\mathcal{K}$ denotes compact operators and $\mathcal{P}$ orthogonal projection. The algebra (3.85) has four irreducible representations. Any element $a \in \mathcal{A}_{P_4(S^1)}$ is of the form

$$a = \lambda \mathcal{P}_{3,12} + k_{34,1} + k_{34,2} + \mu \mathcal{P}_{4,12} , \tag{3.86}$$

with $\lambda, \mu \in \mathbb{C}$, $k_{34,1} \in \mathcal{K}_{\mathcal{H}_{31} \oplus \mathcal{H}_{41}}$ and $k_{34,2} \in \mathcal{K}_{\mathcal{H}_{32} \oplus \mathcal{H}_{42}}$. The representations are the following ones,

$$
\begin{aligned}
\pi_1 &: \mathcal{A}_{P_4(S^1)} \longrightarrow \mathcal{B}(\mathcal{H}) , & a \mapsto \pi_1(a) = \lambda \mathcal{P}_{3,12} + k_{34,1} + \mu \mathcal{P}_{4,12} , \\
\pi_2 &: \mathcal{A}_{P_4(S^1)} \longrightarrow \mathcal{B}(\mathcal{H}) , & a \mapsto \pi_2(a) = \lambda \mathcal{P}_{3,12} + k_{34,2} + \mu \mathcal{P}_{4,12} , \\
\pi_3 &: \mathcal{A}_{P_4(S^1)} \longrightarrow \mathcal{B}(\mathbb{C}) \simeq \mathbb{C} , & a \mapsto \pi_3(a) = \lambda , \\
\pi_4 &: \mathcal{A}_{P_4(S^1)} \longrightarrow \mathcal{B}(\mathbb{C}) \simeq \mathbb{C} , & a \mapsto \pi_4(a) = \mu .
\end{aligned}
\tag{3.87}
$$

The corresponding kernels are

$$
\begin{aligned}
\mathcal{I}_1 &= \mathcal{K}_{\mathcal{H}_{32} \oplus \mathcal{H}_{42}} , \\
\mathcal{I}_2 &= \mathcal{K}_{\mathcal{H}_{31} \oplus \mathcal{H}_{41}} , \\
\mathcal{I}_3 &= \mathcal{K}_{\mathcal{H}_{31} \oplus \mathcal{H}_{41}} + \mathcal{K}_{\mathcal{H}_{32} \oplus \mathcal{H}_{42}} + \mathbb{C}\mathcal{P}_{\mathcal{H}_{41} \oplus \mathcal{H}_{42}} , \\
\mathcal{I}_4 &= \mathbb{C}\mathcal{P}_{\mathcal{H}_{31} \oplus \mathcal{H}_{32}} + \mathcal{K}_{\mathcal{H}_{31} \oplus \mathcal{H}_{41}} + \mathcal{K}_{\mathcal{H}_{32} \oplus \mathcal{H}_{42}} .
\end{aligned}
\tag{3.88}
$$

The partial order given by the inclusions $\mathcal{I}_1 \subset \mathcal{I}_3$, $\mathcal{I}_1 \subset \mathcal{I}_4$ and $\mathcal{I}_2 \subset \mathcal{I}_3$, $\mathcal{I}_2 \subset \mathcal{I}_4$ produces a topological space $Prim\mathcal{A}_{P_4(S^1)}$ which is just the circle poset in Fig. 3.1.

Example 3.4.7. We shall now give an example of a three-level poset. It corresponds to an approximation of a two dimensional topological space (or a portion thereof).

This topological space, shown in Fig. 3.15, contains five closed sets:

$$
\begin{aligned}
K_0 &= \overline{\{x_1\}} = \{x_1, x_2, x_3, x_4\} , \quad K_1 = \overline{\{x_2\}} = \{x_2, x_3, x_4\} , \\
K_2 &= \overline{\{x_3\}} = \{x_3\} , \quad K_3 = \overline{\{x_4\}} = \{x_4\} , \\
K_4 &= \{x_3, x_4\} = K_2 \cup K_3 .
\end{aligned}
\tag{3.89}
$$

Thus, still with the notations of Proposition 3.4.6, one finds,

Fig. 3.15. A poset approximating a two dimensional space

$$
\begin{aligned}
\mathcal{K}_0 &= \{K_0\}\,, & \mathcal{K}'_0 &= \{K_0\}\,, \\
\mathcal{K}_1 &= \{K_0, K_1\}\,, & \mathcal{K}'_1 &= \{K_0, K_1\}\,, \\
\mathcal{K}_2 &= \{K_0, K_1, K_2\}\,, & \mathcal{K}'_2 &= \{K_0, K_1, K_2\}\,, \\
\mathcal{K}_3 &= \{K_0, K_1, K_2, K_3\}\,, & \mathcal{K}'_3 &= \{K_0, K_1, K_2, K_3, K_4\}\,, \\
\mathcal{K}_4 &= \{K_0, K_1, K_2, K_3, K_4\}\,, & \mathcal{K}'_4 &= \{K_0, K_1, K_2, K_3, K_4\}\,, \\
&\ \vdots
\end{aligned}
$$

$$Y_0(1) = \{x_1, x_2, x_3, x_4\}\,, \qquad\qquad F_0(1) = K_0\,,$$

$$
\begin{aligned}
Y_1(1) &= \{x_2, x_3, x_4\}\,, \\
&\qquad Y_1(2) = \{x_1\}\,, & F_1(1) &= K_1\,, & F_1(2) &= K_0\,,
\end{aligned}
$$

$$
\begin{aligned}
Y_2(1) &= \{x_3\}\,, & Y_2(2) &= \{x_1\}\,, & F_2(1) &= K_2\,, & F_2(2) &= K_0\,, \\
Y_2(3) &= \{x_2, x_4\}\,, & & & F_2(3) &= K_1\,,
\end{aligned}
$$

$$
\begin{aligned}
Y_3(1) &= \{x_3\}\,, & Y_3(2) &= \{x_1\}\,, & F_3(1) &= K_2\,, & F_3(2) &= K_0\,, \\
Y_3(3) &= \{x_2\}\,, & Y_3(4) &= \{x_4\}\,, & F_3(3) &= K_1\,, & F_3(4) &= K_3\,,
\end{aligned}
$$

$$
\begin{aligned}
Y_4(1) &= \{x_3\}\,, & Y_4(2) &= \{x_1\}\,, & F_4(1) &= K_2\,, & F_4(2) &= K_0\,, \\
Y_4(3) &= \{x_2\}\,, & Y_4(4) &= \{x_4\}\,, & F_4(3) &= K_1\,, & F_4(4) &= K_3\,,
\end{aligned}
$$

$$\vdots$$

$$(3.90)$$

Since there are a finite number of points (four), and hence a finite number of closed sets (five), the partition of the poset repeats after the fourth level. The corresponding Bratteli diagram is shows in Fig. 3.16. The ideal $\{0\}$ is primitive. The corresponding algebra is given by

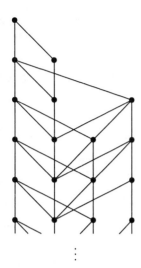

Fig. 3.16. The Bratteli diagram for the poset Y of previous Figure

$$\mathcal{A}_0 = \mathbb{M}_1\,(\mathbb{C})\ ,$$
$$\mathcal{A}_1 = \mathbb{M}_1\,(\mathbb{C}) \oplus \mathbb{M}_1\,(\mathbb{C})\ ,$$
$$\mathcal{A}_2 = \mathbb{M}_1\,(\mathbb{C}) \oplus \mathbb{M}_2\,(\mathbb{C}) \oplus \mathbb{M}_1\,(\mathbb{C})\ ,$$
$$\mathcal{A}_3 = \mathbb{M}_1\,(\mathbb{C}) \oplus \mathbb{M}_4\,(\mathbb{C}) \oplus \mathbb{M}_2\,(\mathbb{C}) \oplus \mathbb{M}_1\,(\mathbb{C})\ ,$$
$$\mathcal{A}_4 = \mathbb{M}_1\,(\mathbb{C}) \oplus \mathbb{M}_8\,(\mathbb{C}) \oplus \mathbb{M}_4\,(\mathbb{C}) \oplus \mathbb{M}_1\,(\mathbb{C})\ , \tag{3.91}$$
$$\vdots$$
$$\mathcal{A}_n = \mathbb{M}_1\,(\mathbb{C}) \oplus \mathbb{M}_{n^2-3n+4}(\mathbb{C}) \oplus \mathbb{M}_{2n-4}\,(\mathbb{C}) \oplus \mathbb{M}_1\,(\mathbb{C})\ ,$$
$$\vdots$$

where, for $n > 2$, $\mathcal{A}_n$ is embedded in $\mathcal{A}_{n+1}$ as follows

$$
\begin{bmatrix} \lambda_1 & & & \\ & B & & \\ & & C & \\ & & & \lambda_2 \end{bmatrix}
\mapsto
\begin{bmatrix}
\lambda_1 & & & & & & & & \\
& \lambda_1 & 0 & 0 & 0 & & & & \\
& 0 & B & 0 & 0 & & & & \\
& 0 & 0 & C & 0 & & & & \\
& 0 & 0 & 0 & \lambda_2 & & & & \\
& & & & & \lambda_1 & 0 & 0 & \\
& & & & & 0 & C & 0 & \\
& & & & & 0 & 0 & \lambda_2 & \\
& & & & & & & & \lambda_2
\end{bmatrix},
$$
$$\tag{3.92}$$

with $\lambda_1, \lambda_2 \in \mathbb{M}_1\,(\mathbb{C})$, $B \in \mathbb{M}_{n^2-3n+4}(\mathbb{C})$ and $C \in \mathbb{M}_{2n-4}\,(\mathbb{C})$; elements which are not shown are equal to zero. Again, the algebra limit $\mathcal{A}_Y$ can be given as a subalgebra of bounded operators on a Hilbert space $\mathcal{H}$. The Hilbert spaces associated with the links of the poset will be $\mathcal{H}_{32}, \mathcal{H}_{42}, \mathcal{H}_{21}$. The difference with the previous example is that now there are links at different levels. On passing from one level to the next (or previous one) one introduces tensor products. The Hilbert space $\mathcal{H}$ is given by

$$\mathcal{H} = \mathcal{H}_{32} \otimes \mathcal{H}_{21} \oplus \mathcal{H}_{42} \otimes \mathcal{H}_{21} \simeq (\mathcal{H}_{32} \oplus \mathcal{H}_{42}) \otimes \mathcal{H}_{21} \ . \qquad (3.93)$$

The algebra $\mathcal{A}_Y$ is then given by [61],

$$\mathcal{A}_Y = \mathbb{C}P_{\mathcal{H}_{32} \otimes \mathcal{H}_{21}} + \mathcal{K}_{\mathcal{H}_{32} \oplus \mathcal{H}_{42}} \otimes P_{\mathcal{H}_{21}} + \mathcal{K}_{(\mathcal{H}_{32} \oplus \mathcal{H}_{42}) \otimes \mathcal{H}_{21}} + \mathbb{C}P_{\mathcal{H}_{42} \otimes \mathcal{H}_{21}} \ . \qquad (3.94)$$

Here $\mathcal{K}$ denotes compact operators and P orthogonal projection. This algebra has four irreducible representations. Any element of it is of the form

$$a = \lambda P_{321} + k_{34,2} \otimes P_{21} + k_{34,21} + \mu P_{421} \ , \qquad (3.95)$$

with $\lambda, \mu \in \mathbb{C}$, $k_{34,2} \in \mathcal{K}_{\mathcal{H}_{32} \oplus \mathcal{H}_{42}}$ and $k_{34,21} \in \mathcal{K}_{(\mathcal{H}_{32} \oplus \mathcal{H}_{42}) \otimes \mathcal{H}_{21}}$. The representations are the following ones,

$$
\begin{aligned}
\pi_1 &: \mathcal{A}_Y \longrightarrow \mathcal{B}(\mathcal{H}) \ , & a &\mapsto \pi_1(a) = \lambda P_{321} + k_{34,2} \otimes P_{21} + k_{34,21} + \mu P_{421} \ , \\
\pi_2 &: \mathcal{A}_Y \longrightarrow \mathcal{B}(\mathcal{H}) \ , & a &\mapsto \pi_2(a) = \lambda P_{321} + k_{34,2} \otimes P_{21} + \mu P_{421} \ , \\
\pi_3 &: \mathcal{A}_Y \longrightarrow \mathcal{B}(\mathbb{C}) \simeq \mathbb{C} \ , & a &\mapsto \pi_3(a) = \lambda \ , \\
\pi_4 &: \mathcal{A}_Y \longrightarrow \mathcal{B}(\mathbb{C}) \simeq \mathbb{C} \ , & a &\mapsto \pi_4(a) = \mu \ .
\end{aligned}
$$
$$\qquad (3.96)$$

The corresponding kernels are

$$
\begin{aligned}
\mathcal{I}_1 &= \{0\} \ , \\
\mathcal{I}_2 &= \mathcal{K}_{(\mathcal{H}_{32} \oplus \mathcal{H}_{42}) \otimes \mathcal{H}_{21}} \ , \\
\mathcal{I}_3 &= \mathcal{K}_{\mathcal{H}_{32} \oplus \mathcal{H}_{42}} \otimes P_{\mathcal{H}_{21}} + \mathcal{K}_{(\mathcal{H}_{32} \oplus \mathcal{H}_{42}) \otimes \mathcal{H}_{21}} + \mathbb{C}P_{\mathcal{H}_{42} \otimes \mathcal{H}_{21}} \ , \\
\mathcal{I}_4 &= \mathbb{C}P_{\mathcal{H}_{32} \otimes \mathcal{H}_{21}} + \mathcal{K}_{\mathcal{H}_{32} \oplus \mathcal{H}_{42}} \otimes P_{\mathcal{H}_{21}} + \mathcal{K}_{(\mathcal{H}_{32} \oplus \mathcal{H}_{42}) \otimes \mathcal{H}_{21}} \ .
\end{aligned}
$$
$$\qquad (3.97)$$

The partial order given by the inclusions $\mathcal{I}_1 \subset \mathcal{I}_2 \subset \mathcal{I}_3$ and $\mathcal{I}_1 \subset \mathcal{I}_2 \subset \mathcal{I}_4$ produces a topological space $Prim\mathcal{A}_Y$ which is just the poset of Fig. (3.15).

In fact, by looking at the previous examples a bit more carefully one can infer the algorithm by which one goes from a (finite) poset P to the corresponding Bratteli diagram $\mathcal{D}(\mathcal{A}_P)$. Let $(x_1, \cdots, x_N)$ be the points of P and for $k = 1, \cdots, N$, let $S_k =: \overline{\{x_k\}}$ be the smallest closed subset of P containing the point x_j. Then, the Bratteli diagram repeats itself after level N and the partition $Y_n(k)$ of Proposition 3.4.6 is just given by

$$Y_n(k) = Y_{n+1}(k) = \{x_k\} \ , \quad k = 1, \ldots, N \ , \quad \forall\, n \geq N \ . \qquad (3.98)$$

As for the associated $F_n(k)$ of the same Proposition, from level $N + 1$ on, they are given by the S_k,

$$F_n(k) = F_{n+1}(k) = S_k \ , \quad k = 1, \ldots, N \ , \quad \forall\, n \geq N + 1 \ . \qquad (3.99)$$

In the diagram $\mathcal{D}(\mathcal{A}_P)$, for any $n \geq N$, $(n, k) \searrow (n + 1, j)$ if and only if $\overline{\{x_k\}} \cap S_j \neq \emptyset$, that is if and only if $x_k \in S_j$.

We also sketch the algorithm used to construct the algebra limit $\mathcal{A}_P$ determined by the Bratteli diagram $\mathcal{D}(\mathcal{A}_P)^{14}$ [8, 61]. The idea is to associate to the poset P an infinite dimensional separable Hilbert space $\mathcal{H}(P)$ out of tensor products and direct sums of infinite dimensional (separable) Hilbert spaces $\mathcal{H}_{ij}$ associated with each link $(x_i, x_j), x_i \succ x_j$, in the poset.[15] Then for each point $x \in P$ there is a subspace $\mathcal{H}(x) \subset \mathcal{H}(P)$ and an algebra $\mathcal{B}(x)$ of bounded operators acting on $\mathcal{H}(x)$. The algebra $\mathcal{A}_P$ is the one generated by all the $\mathcal{B}(x)$ as x varies in P. In fact, the algebra $\mathcal{B}(x)$ can be made to act on the whole of $\mathcal{H}(P)$ by defining its action on the complement of $\mathcal{H}(x)$ to be zero. Consider any *maximal chain* C_α in P: $C_\alpha = \{x_\alpha, \ldots, x_2, x_1 \mid x_j \succ x_{j-1}\}$ for any maximal point $x_\alpha \in P$. To this chain one associates the Hilbert space

$$\mathcal{H}(C_\alpha) = \mathcal{H}_{\alpha,\alpha-1} \otimes \cdots \otimes \mathcal{H}_{3,2} \otimes \mathcal{H}_{2,1} . \tag{3.100}$$

By taking the direct sum over all maximal chains, one gets the Hilbert space $\mathcal{H}(P)$,

$$\mathcal{H}(P) = \bigoplus_\alpha \mathcal{H}(C_\alpha) . \tag{3.101}$$

The subspace $\mathcal{H}(x) \subset \mathcal{H}(P)$ associated with any point $x \in P$ is constructed in a similar way by restricting the sum to all maximal chains containing the point x. It can be split into two parts,

$$\mathcal{H}(x) = \mathcal{H}(x)^u \otimes \mathcal{H}(x)^d , \tag{3.102}$$

with,

$$\begin{aligned} \mathcal{H}(x)^u &= \mathcal{H}(P_x^u) , & P_x^u &= \{y \in P \mid y \succeq x\} , \\ \mathcal{H}(x)^d &= \mathcal{H}(P_x^d) , & P_x^d &= \{y \in P \mid y \preceq x\} . \end{aligned} \tag{3.103}$$

Here $\mathcal{H}(P_x^u)$ and $\mathcal{H}(P_x^d)$ are constructed as in (3.101); also, $\mathcal{H}(x)^u = \mathbb{C}$ if x is a maximal point and $\mathcal{H}(x)^d = \mathbb{C}$ if x is a minimal point. Consider now the algebra $\mathcal{B}(x)$ of bounded operators on $\mathcal{H}(x)$ given by

$$\mathcal{B}(x) = \mathcal{K}(\mathcal{H}(x)^u) \otimes \mathbb{C}P(\mathcal{H}(x)^d) \simeq \mathcal{K}(\mathcal{H}(x)^u) \otimes P(\mathcal{H}(x)^d) . \tag{3.104}$$

As before, $\mathcal{K}$ denotes compact operators and $\mathcal{P}$ orthogonal projection. We see that $\mathcal{B}(x)$ acts by compact operators on the Hilbert space $\mathcal{H}(x)^u$ determined by the points which follow x and by multiples of the identity on the Hilbert space $\mathcal{H}(x)^d$ determined by the points which precede x. These algebras satisfy the rules: $\mathcal{B}(x)\mathcal{B}(y) \subset \mathcal{B}(x)$ if $x \preceq y$ and $\mathcal{B}(x)\mathcal{B}(y) = 0$ if x and y are not comparable. As already mentioned, the algebra $\mathcal{A}(P)$ of the poset P is the algebra of bounded operators on $\mathcal{H}(P)$ generated by all $\mathcal{B}(x)$ as x varies over P. It can be shown that $\mathcal{A}(P)$ has a space of primitive ideals which is homeomorphic to the poset P [8, 61]. We refer to [60, 61] for additional details and examples.

---

[14] This algebra is really defined only modulo Morita equivalence.

[15] The Hilbert spaces could all be taken to be the same. The label is there just to distinguish among them.

3.5 How to Recover the Algebra Being Approximated

In Sect. 3.3 we have described how to recover a topological space M in the limit, by considering a sequence of finer and finer coverings of M. We constructed a projective system of finitary topological spaces and continuous maps $\{P_i, \pi_{ij}\}_{i,j \in \mathbb{N}}$ associated with the coverings; the maps $\pi_{ij} : P_j \to P_i$, $j \geq i$, being continuous surjections. The limit of the system is a topological space P_∞, in which M is embedded as the subspace of closed points. On each point m of (the image of) M there is a fiber of 'extra points'; the latter are all points of P_∞ which 'cannot be separated' by m.

Well, from a dual point of view we get a *inductive system of algebras and homomorphisms* $\{\mathcal{A}_i, \phi_{ij}\}_{i,j \in \mathbb{N}}$; the maps $\phi_{ij} : \mathcal{A}_i \to \mathcal{A}_j$, $j \geq i$, being injective homeomorphisms. The system has a unique *inductive limit* $\mathcal{A}^\infty$. Each algebra $\mathcal{A}_i$ is such that $\widehat{\mathcal{A}_i} = P_i$ and is associated with P_i as described previously, $\mathcal{A}_i = \mathcal{A}(P_i)$. The map ϕ_{ij} is a 'suitable pullback' of the corresponding surjection π_{ij}. The limit space P_∞ is the structure space of the limit algebra $\mathcal{A}^\infty$, $P_\infty = \widehat{\mathcal{A}^\infty}$. And, finally the algebra $C(M)$ of continuous functions on M can be identified with the *center* of $\mathcal{A}^\infty$.

We also get a inductive system of Hilbert spaces together with isometries $\{\mathcal{H}_i, \tau_{ij}\}_{i,j \in \mathbb{N}}$; the maps $\tau_{ij} : \mathcal{H}_i \to \mathcal{H}_j$, $j \geq i$, being injective isometries onto the image. The system has a unique *inductive limit* $\mathcal{H}^\infty$. Each Hilbert space $\mathcal{H}_i$ is associated with the space P_i as in (3.101), $\mathcal{H}_i = \mathcal{H}(P_i)$, the algebra $\mathcal{A}_i$ being the corresponding subalgebra of bounded operators. The maps τ_{ij} are constructed out of the corresponding ϕ_{ij}. The limit Hilbert space $\mathcal{H}^\infty$ is associated with the space P_∞ as in (3.101), $\mathcal{H}^\infty = \mathcal{H}(P_\infty)$, the algebra $\mathcal{A}^\infty$ again being the corresponding subalgebra of bounded operators. And, finally, the Hilbert space $L^2(M)$ of square integrable functions is 'contained' in $\mathcal{H}^\infty$: $\mathcal{H}^\infty = L^2(M) \oplus_\alpha \mathcal{H}_\alpha$, the sum being on the 'extra points' in P_∞.

All of the previous is described in great details in [11]. Here we only make a few additional remarks. By improving the approximation (by increasing the number of 'detectors') one gets a noncommutative lattice whose Hasse diagram has a bigger number of points and links. The associated Hilbert space gets 'more refined' : one may think of a *unique (and the same)* Hilbert space which is being refined while being split by means of tensor products and direct sums. In the limit the information is enough to recover completely the Hilbert space (in fact, to recover more than it). Further considerations along these lines and possible applications to quantum mechanics will have to await another occasion.

3.6 Operator Valued Functions on Noncommutative Lattices

Much in the same way as it happens for the commutative algebras described in Sect. 2.2, elements of a noncommutative C^*-algebra whose primitive spectrum $Prim\mathcal{A}$ is a noncommutative lattice can be realized as operator-valued functions on $Prim\mathcal{A}$. The value of $a \in \mathcal{A}$ at the 'point' $\mathcal{I} \in Prim\mathcal{A}$ is just the image of a under the representation $\pi_\mathcal{I}$ associated with $\mathcal{I}$ and such that $\ker(\pi_\mathcal{I}) = \mathcal{I}$,

$$a(\mathcal{I}) = \pi_\mathcal{I}(a) \simeq a/\mathcal{I} , \quad \forall\, a \in \mathcal{A},\ \mathcal{I} \in Prim\mathcal{A} . \qquad (3.105)$$

All this is shown pictorially in Figs. 3.17, 3.18 and 3.19 for the $\bigvee$ lattice,

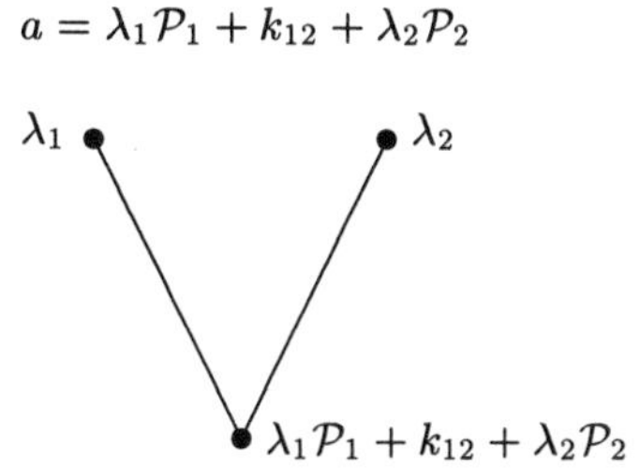

$a = \lambda_1 \mathcal{P}_1 + k_{12} + \lambda_2 \mathcal{P}_2$

$\lambda_1 \bullet \qquad\qquad \bullet\, \lambda_2$

$\bullet\, \lambda_1 \mathcal{P}_1 + k_{12} + \lambda_2 \mathcal{P}_2$

Fig. 3.17. A function over the lattice $\bigvee$

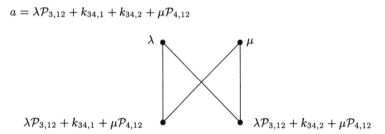

$a = \lambda \mathcal{P}_{3,12} + k_{34,1} + k_{34,2} + \mu \mathcal{P}_{4,12}$

$\lambda \bullet \qquad\qquad \bullet\, \mu$

$\lambda \mathcal{P}_{3,12} + k_{34,1} + \mu \mathcal{P}_{4,12} \bullet \qquad\qquad \bullet\, \lambda \mathcal{P}_{3,12} + k_{34,2} + \mu \mathcal{P}_{4,12}$

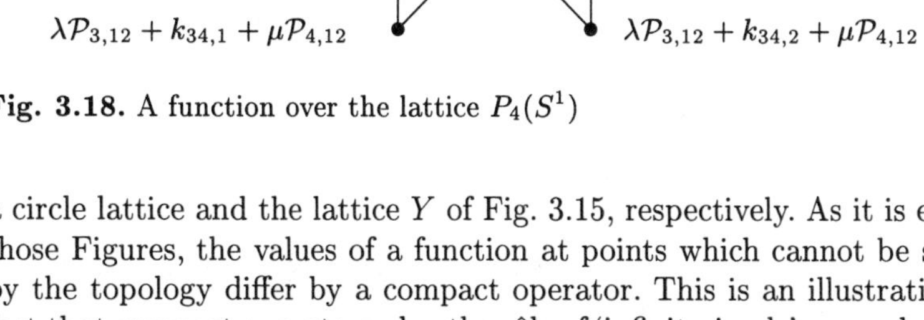

Fig. 3.18. A function over the lattice $P_4(S^1)$

a circle lattice and the lattice Y of Fig. 3.15, respectively. As it is evident in those Figures, the values of a function at points which cannot be separated by the topology differ by a compact operator. This is an illustration of the fact that compact operators play the rôle of 'infinitesimals' as we shall discuss at length in Sect. 6.1. Furthermore, while in Figs. 3.17 and 3.18 we have only 'infinitesimals of first order', for the three level lattice of Fig. 3.19 we have both infinitesimals of first order, like $k_{34,2}$, and infinitesimals of second order, like $k_{34,21}$.

In fact, as we shall see in Sect. 4.2, the correct way of thinking of any noncommutative C^*-algebra $\mathcal{A}$ is as the module of sections of the 'rank one

$$a = \lambda P_{321} + k_{34,2} \otimes P_{21} + k_{34,21} + \mu P_{421}$$

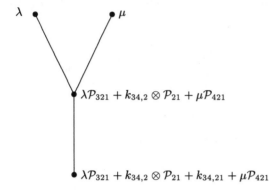

Fig. 3.19. A function over the lattice Y

trivial vector bundle' over the associated noncommutative space. For the kind of noncommutative lattices we are interested in, it is possible to explicitly construct the bundle over the lattice. Such bundles are examples of *bundles of C*-algebras* [55], the fiber over any point $\mathcal{I} \in Prim\mathcal{A}$ being just the algebra of bounded operators $\pi_{\mathcal{I}}(\mathcal{A}) \subset \mathcal{B}(\mathcal{H}_{\mathcal{I}})$, with $\mathcal{H}_{\mathcal{I}}$ the representation space. The Hilbert space and the algebra are given explicitly by the Hilbert space in (3.102) and the algebra in (3.104) respectively, by taking for x the point $\mathcal{I}$.[16] It is also possible to endow the total space with a topology in such a manner that elements of $\mathcal{A}$ are realized as continuous sections. We refer to [62] for more details. Figure 3.20 shows the trivial bundle over the lattice $P_4(S^1)$.

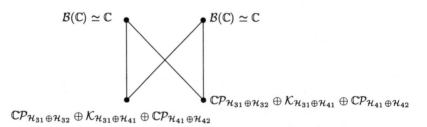

Fig. 3.20. The fibers of the trivial bundle over the lattice $P_4(S^1)$

---

[16] At the same time, one is also constructing a bundle of Hilbert spaces.

4. Modules as Bundles

The algebraic analogue of vector bundles has its origin in the fact that a vector bundle $E \to M$ over a manifold M is completely characterized by the space $\mathcal{E} = \Gamma(E, M)$ of its smooth sections. In this context, the space of sections is thought of as a (right) module over the algebra $C^\infty(M)$ of smooth functions over M. Indeed, by the Serre-Swan theorem [123], locally trivial, finite-dimensional complex vector bundles over a compact Hausdorff space M correspond canonically to finite projective modules over the algebra $\mathcal{A} = C^\infty(M)$.[1] To the vector bundle E one associates the $C^\infty(M)$-module $\mathcal{E} = \Gamma(M, E)$ of smooth sections of E. Conversely, if $\mathcal{E}$ is a finite projective module over $C^\infty(M)$, the fiber E_m of the associated bundle E over the point $m \in M$ is

$$E_m = \mathcal{E}/\mathcal{E}\mathcal{I}_m \, , \tag{4.1}$$

where the ideal $\mathcal{I}_m \subset \mathcal{C}(M)$, corresponding to the point $m \in M$, is given by

$$
\begin{aligned}
\mathcal{I}_m &= \ker\{\chi_m : C^\infty(M) \to \mathbb{C} \mid \chi_m(f) = f(m)\} \\
&= \{f \in C^\infty(M) \mid f(m) = 0\} \, .
\end{aligned}
\tag{4.2}
$$

However, given a noncommutative algebra $\mathcal{A}$, playing the rôle of the algebra of smooth functions on some noncommutative space, one has more than one possibility for the analogue of a vector bundle. The possible relevant categories seem to be the following ones [49],

1. left or right $\mathcal{A}$-modules;
2. $\mathcal{A}$-bimodules of a particular kind (see later);
3. modules over the center $Z(\mathcal{A})$ of $\mathcal{A}$.

We shall start by describing right (and left) *projective modules of finite type* (or *finite projective modules*) over $\mathcal{A}$. These classes of modules provide suitable substitutes for the analogue of complex vector bundles, although one has to introduce additional structures in order to consider tensor products. Hermitian vector bundles, namely bundles with a Hermitian structure, correspond to projective modules of finite type $\mathcal{E}$ endowed with an $\mathcal{A}$-valued

---

[1] In fact, in [123] the correspondence is stated in the continuous category, meaning for functions and sections which are continuous. However, it can be extended to the smooth case, see [33].

sesquilinear form. For A a C^*-algebra, the appropriate notion is that of Hilbert module that we describe at length in App. A.3.

We start with some machinery from the theory of modules which we take mainly from [20].

4.1 Modules

Definition 4.1.1. *Suppose we are given an algebra A over (say) the complex numbers $\mathbb{C}$. A vector space $\mathcal{E}$ over $\mathbb{C}$ is also a right module over A if it carries a right representation of A,*

$$\mathcal{E} \times A \ni (\eta, a) \mapsto \eta a \in \mathcal{E} , \qquad \eta(ab) = (\eta a)b ,$$
$$\eta(a + b) = \eta a + \eta b ,$$
$$(\eta + \xi)a = \eta a + \xi a , \qquad (4.3)$$

for any $\eta, \xi \in \mathcal{E}$ and $a, b \in A$.

Definition 4.1.2. *Given two right A-modules $\mathcal{E}$ and $\mathcal{F}$, a morphism of $\mathcal{E}$ into $\mathcal{F}$ is any linear map $\rho : \mathcal{E} \to \mathcal{F}$ which in addition is A-linear,*

$$\rho(\eta a) = \rho(\eta)a , \quad \forall \, \eta \in \mathcal{E}, \, a \in A . \qquad (4.4)$$

A *left module* and a *morphism of left modules* are defined in a similar way. In general, a right module structure and a left module one should be taken to be distinct.

A *bimodule* over the algebra A is a vector space $\mathcal{E}$ which carries both a left and a right A-module structure. Furthermore, the two structures are required to be compatible, namely

$$(a\eta)b = a(\eta b) , \quad \forall \, \eta \in \mathcal{E} , \, a, b \in A . \qquad (4.5)$$

Given a right A-module $\mathcal{E}$, its *dual* $\mathcal{E}'$ is defined as the collection of all morphisms from $\mathcal{E}$ into A,

$$\mathcal{E}' = Hom_A(\mathcal{E}, A) =: \{\phi : \mathcal{E} \to A \mid \phi(\eta a) = \phi(\eta)a , \; \eta \in \mathcal{E}, \, a \in A\} . \quad (4.6)$$

By using the canonical bimodule structure of A, the collection $\mathcal{E}'$ is endowed with a natural *left* A-module structure,

$$A \times \mathcal{E}' \ni (a, \phi) \mapsto a \cdot \phi \in \mathcal{E}' , \quad a \cdot \phi(\eta) =: a(\phi(\eta)) , \qquad (4.7)$$

for $a \in A$, $\phi \in \mathcal{E}'$, $\eta \in \mathcal{E}$. Left analogues of the requirements (4.3) are easily proven.

For any algebra A, the *opposite algebra* A^o has elements a^o in bijective correspondence with the elements $a \in A$ while multiplication is given by $a^o b^o = (ba)^o$. Any right (respectively left) A-module $\mathcal{E}$ can be regarded as a left (respectively right) A^o-module by setting $a^o \eta = \eta a$ (respectively $a\eta =$

ηa^o), for any $\eta \in \mathcal{E}, a \in \mathcal{A}$. The algebra $\mathcal{A}^e =: \mathcal{A} \otimes_{\mathbb{C}} \mathcal{A}^o$ is called the *enveloping algebra* of $\mathcal{A}$. Any $\mathcal{A}$-bimodule $\mathcal{E}$ can be regarded as a right $\mathcal{A}^e$-module by setting $\eta(a \otimes b^o) = b\eta a$, for any $\eta \in \mathcal{E}, a \in \mathcal{A}, b^o \in \mathcal{A}^o$. One can also regard $\mathcal{E}$ as a left $\mathcal{A}^e$-module by setting $(a \otimes b^o)\eta = a\eta b$, for any $\eta \in \mathcal{E}, a \in \mathcal{A}, b^o \in \mathcal{A}^o$.

If $Z(\mathcal{A})$ is the center of $\mathcal{A}$, one can define module structures exactly as before with the rôle of $\mathcal{A}$ taken over by the commutative algebra $Z(\mathcal{A})$. An $\mathcal{A}$-module structure produces a similar one over $Z(\mathcal{A})$, but the converse is clearly not true. Also, one should notice that in spite of the commutativity of $Z(\mathcal{A})$, a right $Z(\mathcal{A})$-module structure and a left $Z(\mathcal{A})$-module structure should again be taken to be distinct.

A family $(e_\lambda)_{\lambda \in \Lambda}$, with Λ any (finite or infinite) directed set, is called a *generating family* for the right module $\mathcal{E}$ if any element of $\mathcal{E}$ can be written (possibly in more than one way) as a combination $\sum_{\lambda \in \Lambda} e_\lambda a_\lambda$, with $a_\lambda \in \mathcal{A}$ and such that only a finite number of terms in the sum are different from zero. The family $(e_\lambda)_{\lambda \in \Lambda}$ is called *free* if it is made up of linearly independent elements (over $\mathcal{A}$), and it is a *basis* for the module $\mathcal{E}$ if it is a free generating family, so that any $\eta \in \mathcal{E}$ can be written uniquely as a combination $\sum_{\lambda \in \Lambda} e_\lambda a_\lambda$, with $a_\lambda \in \mathcal{A}$. The module is called *free* if it admits a basis. The module $\mathcal{E}$ is said to be of *finite type* if it is finitely generated, namely if it admits a generating family of finite cardinality.

Consider now the module $\mathbb{C}^N \otimes_{\mathbb{C}} \mathcal{A} =: \mathcal{A}^N$. Any of its elements η can be thought of as an N-dimensional vector with entries in $\mathcal{A}$ and can be written uniquely as a linear combination $\eta = \sum_{j=1}^{N} e_j a_j$, with $a_j \in \mathcal{A}$ and the basis $\{e_j, j = 1, \ldots, N\}$ being identified with the canonical basis of $\mathbb{C}^N$. This module is clearly both free and of finite type.

A general free module (of finite type) *might* admit bases of different cardinality and so it does not make sense to talk of dimension. If the free module is such that any two bases have the same cardinality, the latter is called the *dimension* of the module.[2]

However, if the module $\mathcal{E}$ is of finite type there is always an integer N and a (module) surjection $\rho : \mathcal{A}^N \to \mathcal{E}$. In this case one can find a basis $\{\epsilon_j, j = 1, \ldots, N\}$ which is just the image of the free basis of $\mathcal{A}^N$, $\epsilon_j = \rho(e_j)$, $j = 1, \ldots, N$. Notice that in general it is not possible to solve the constraints among the basis elements so as to get a free basis. For example, consider the algebra $C^\infty(S^2)$ of smooth functions on the two-dimensional sphere S^2 and the Lie algebra $\mathcal{X}(S^2)$ of smooth vector fields on S^2. Then, $\mathcal{X}(S^2)$ is a module of finite type over $C^\infty(S^2)$, a basis of three elements being given by $\{Y_i = \sum_{j,k=1}^{3} \varepsilon_{ijk} x_k \frac{\partial}{\partial x_k}, i = 1, 2, 3\}$ with x_1, x_2, x_3, such that $\sum_{j}^{3}(x_j)^2 = 1$ (the x_i's are just the natural coordinates of S^2). The basis is not free, since $\sum_{j=1}^{3} x_j Y_j = 0$; there do not exist two globally defined vector

---

[2] A sufficient condition for this to happen is the existence of a (ring) homomorphism $\rho : \mathcal{A} \to \mathbb{D}$, with $\mathbb{D}$ any field. This is, for instance, the case if $\mathcal{A}$ is commutative (since then $\mathcal{A}$ admits at least a maximal ideal M and $\mathcal{A}/M$ is a field) or if $\mathcal{A}$ may be written as a (ring) direct sum $\mathcal{A} = \mathbb{C} \oplus \overline{\mathcal{A}}$ [21].

fields on S^2 which could serve as a basis of $\mathcal{X}(S^2)$. Of course this is nothing but the statement that the tangent bundle TS^2 over S^2 is non-trivial.

4.2 Projective Modules of Finite Type

Definition 4.2.1. *A right $\mathcal{A}$-module $\mathcal{E}$ is said to be* projective *if it satisfies one of the following equivalent properties:*

(1.) (Lifting property.) Given a surjective homomorphism $\rho : \mathcal{M} \to \mathcal{N}$ of right $\mathcal{A}$-modules, any homomorphism $\lambda : \mathcal{E} \to \mathcal{N}$ can be lifted to a homomorphism $\widetilde{\lambda} : \mathcal{E} \to \mathcal{M}$ such that $\rho \circ \widetilde{\lambda} = \lambda$,

$$id: \quad \mathcal{M} \quad \longleftrightarrow \quad \mathcal{M}$$

$$\widetilde{\lambda} \uparrow \qquad \qquad \downarrow \rho$$

$$\lambda: \quad \mathcal{E} \quad \longrightarrow \quad \mathcal{N} \quad , \qquad \rho \circ \widetilde{\lambda} = \lambda \; . \tag{4.8}$$

$$\downarrow$$

$$0$$

(2.) Every surjective module morphism $\rho : \mathcal{M} \to \mathcal{E}$ splits, namely there exists a module morphism $s : \mathcal{E} \to \mathcal{M}$ such that $\rho \circ s = id_\mathcal{E}$.

(3.) The module $\mathcal{E}$ is a direct summand of a free module, that is there exists a free module $\mathcal{F}$ and a module $\mathcal{E}'$ (which will then be projective as well), such that

$$\mathcal{F} = \mathcal{E} \oplus \mathcal{E}' \; . \tag{4.9}$$

To prove that (1.) implies (2.) it is enough to apply (1.) to the case $\mathcal{N} = \mathcal{E}$, $\lambda = id_\mathcal{E}$, and identify $\widetilde{\lambda}$ with the splitting map s. To prove that (2.) implies (3.) one first observes that (2.) implies that $\mathcal{E}$ is a direct summand of $\mathcal{M}$ through s, $\mathcal{M} = s(\mathcal{E}) \oplus \ker\rho$. Also, as mentioned before, for any module $\mathcal{E}$ it is possible to construct a surjection from a free module $\mathcal{F}$, $\rho : \mathcal{F} \to \mathcal{E}$ (in fact $\mathcal{F} = \mathcal{A}^N$ for some N). One then applies (2.) to this surjection. To prove that (3.) implies (1.) one observes that a free module is projective and that a direct sum of modules is projective if and only if any summand is.

Suppose now that $\mathcal{E}$ is both projective and of finite type with surjection $\rho : \mathcal{A}^N \to \mathcal{E}$. Then, the projective properties of Definition 4.2.1 allow one to find a lift $\widetilde{\lambda} : \mathcal{E} \to \mathcal{A}^N$ such that $\rho \circ \widetilde{\lambda} = id_\mathcal{E}$,

$$id: \quad \mathcal{A}^N \quad \longleftrightarrow \quad \mathcal{A}^N$$

$$\widetilde{\lambda} \uparrow \qquad \qquad \downarrow \rho \quad , \qquad \rho \circ \widetilde{\lambda} = id_\mathcal{E} \; . \tag{4.10}$$

$$id: \quad \mathcal{E} \quad \longrightarrow \quad \mathcal{E}$$

We can then construct an idempotent $p \in End_{\mathcal{A}} \mathcal{A}^N \simeq \mathbb{M}_N(\mathcal{A})$, $\mathbb{M}_N(\mathcal{A})$ being the algebra of $N \times N$ matrices with entries in $\mathcal{A}$, given by

$$p = \widetilde{\lambda} \circ \rho \ . \tag{4.11}$$

Indeed, from (4.10), $p^2 = \widetilde{\lambda} \circ \rho \circ \widetilde{\lambda} \circ \rho = \widetilde{\lambda} \circ \rho = p$. The idempotent p allows one to decompose the free module $\mathcal{A}^N$ as a direct sum of submodules,

$$\mathcal{A}^N = p\mathcal{A}^N \oplus (1-p)\mathcal{A}^N \tag{4.12}$$

and in this way ρ and $\widetilde{\lambda}$ are isomorphisms (one the inverse of the other) between $\mathcal{E}$ and $p\mathcal{A}^N$. The module $\mathcal{E}$ is then projective of finite type over $\mathcal{A}$ if and only if there exits an idempotent $p \in \mathbb{M}_N(\mathcal{A})$, $p^2 = p$, such that $\mathcal{E} = p\mathcal{A}^N$. We may think of elements of $\mathcal{E}$ as N-dimensional column vectors whose elements are in $\mathcal{A}$, the collection of which are invariant under the action of p,

$$\mathcal{E} = \{\xi = (\xi_1, \ldots, \xi_N) \ ; \ \xi_i \in \mathcal{A} \ , \ p\xi = \xi\} \ . \tag{4.13}$$

In the what follows, we shall use the term *finite projective* to mean *projective of finite type*.

The crucial link between finite projective modules and vector bundles is provided by the following central result which is named after Serre and Swan [123] (see also [130]). As mentioned before, the Serre-Swan theorem was established for functions and sections which are continuous; it can however, be extended to the smooth case [33].

Proposition 4.2.1. *Let M be a compact finite dimensional manifold. A $C^\infty(M)$-module $\mathcal{E}$ is isomorphic to a module $\Gamma(E, M)$ of smooth sections of a bundle $E \to M$, if and only if it is finite projective.*

Proof. We first prove that a module $\Gamma(E, M)$ of sections is finite projective. If $E \simeq M \times \mathbb{C}^k$ is the rank k trivial vector bundle, then $\Gamma(E, M)$ is just the free module $\mathcal{A}^k$, $\mathcal{A}$ being the algebra $C^\infty(M)$. In general, from what was said before, one has to construct two maps $\lambda : \Gamma(E, M) \to \mathcal{A}^N$ (this was called $\widetilde{\lambda}$ before), and $\rho : \mathcal{A}^N \to \Gamma(E, M), N$ being a suitable integer, such that $\rho \circ \lambda = id_{\Gamma(E,M)}$. Then $\Gamma(E, M) = p\mathcal{A}^N$, with the idempotent p given by $p = \lambda \circ \rho$. Let $\{U_i, i = 1, \cdots, q\}$ be an open covering of M. Any element $s \in \Gamma(E, M)$ can be represented by q smooth maps $s_i = s_{|U_i} : U \to \mathbb{C}^k$, which satisfy the compatibility conditions

$$s_j(m) = \sum_j g_{ji}(m)s_i(m) \ , \quad m \in U_i \cap U_j \ , \tag{4.14}$$

with $g_{ji} : U_i \cap U_j \to GL(k, \mathbb{C})$ the transition functions of the bundle. Consider now a partition of unity $\{h_i, , i = 1, \cdots, q\}$ subordinate to the covering $\{U_i\}$. By a suitable rescaling we can always assume that $h_1^2 + \cdots + h_q^2 = 1$ so that h_j^2 is a partition of unity subordinate to $\{U_i\}$ as well. Now set $N = kq$, write $\mathbb{C}^N = \mathbb{C}^k \oplus \cdots \oplus \mathbb{C}^k$ (q summands), and define

$$\lambda : \Gamma(E, M) \to \mathcal{A}^N \ ,$$
$$\lambda(s_1, \cdots, s_q) =: (h_1 s_1, \cdots, h_q s_q) \ ,$$

$$\rho : \mathcal{A}^N \to \Gamma(E, M) \ ;$$
$$\rho(t_1, \cdots, t_q) =: (\tilde{s}_1, \cdots, \tilde{s}_q) \ , \quad \tilde{s}_i = \sum_j g_{ij} h_j t_j \ . \tag{4.15}$$

Then

$$\rho \circ \lambda(s_1, \cdots, s_q) = (\tilde{s}_1, \cdots, \tilde{s}_q) \ , \quad \tilde{s}_i = \sum_j g_{ij} h_j^2 s_j \ , \tag{4.16}$$

which, since $\{h_j^2\}$ is a partition of unity, amounts to $\rho \circ \lambda = id_{\Gamma(E,M)}$.

Conversely, suppose that $\mathcal{E}$ is a finite projective $C^\infty(M)$-module. Then, with $\mathcal{A} = C^\infty(M)$, one can find an integer N and an idempotent $p \in \mathbb{M}_N(\mathcal{A})$, such that $\mathcal{E} = p\mathcal{A}^N$. Now, $\mathcal{A}^N$ can be identified with the module of sections of the trivial bundle $M \times \mathbb{C}^N$, $\mathcal{A}^N \simeq \Gamma(M \times \mathbb{C}^N)$. Since p is a module map, one has that $p(sf) = p(s)f$, $f \in C^\infty(M)$. If $m \in M$ and $\mathcal{I}_m$ is the ideal $\mathcal{I}_m = \{f \in C^\infty(M) \mid f(m) = 0\}$, then p preserves the submodule $\mathcal{A}^N \mathcal{I}_m$. Since $s \mapsto s(m)$ induces a linear isomorphism of $\mathcal{A}^N / \mathcal{A}^N \mathcal{I}_m$ onto the fiber $(M \times \mathbb{C}^N)_m$, we have that $p(s)(m) \in (M \times \mathbb{C}^N)_m$ for all $s \in \mathcal{A}^N$. Then the map $\pi : M \times \mathbb{C}^N \to M \times \mathbb{C}^N$, $s(m) \mapsto p(s)(m)$, defines a bundle homomorphism satisfying $p(s) = \pi \circ s$. Since $p^2 = p$, one has that $\pi^2 = \pi$. Suppose now that $dim \ \pi((M \times \mathbb{C}^N)_m) = k$. Then one can find k linearly independent smooth local sections $s_j \in \mathcal{A}^N$, $j = 1, \cdots, k$, near $m \in M$, such that $\pi \circ s_j(m) = s_j(m)$. Then, $\pi \circ s_j$, $j = 1, \cdots, k$ are linearly independent in a neighborhood U of m, so that $dim \ \pi((M \times \mathbb{C}^N)_{m'}) \geq k$, for any $m' \in U$. Similarly, by considering the idempotent $(1 - \pi) : M \times \mathbb{C}^N \to M \times \mathbb{C}^N$, one gets that $dim \ (1 - \pi)((M \times \mathbb{C}^N)_{m'}) \geq N - k$, for any $m' \in U$. The integer N being constant, one infers that $dim \ \pi((M \times \mathbb{C}^N)_{m'})$ is (locally) constant, so that $\pi(M \times \mathbb{C}^N)$ is the total space of a vector bundle $E \to M$ which is such that $M \times \mathbb{C}^N = E \oplus \ker \pi$. From the way the bundle E is constructed, the module of its sections is given by $\Gamma(E, M) = \{\pi \circ s \mid s \in \Gamma(M \times \mathbb{C}^N)\} = Im\{p : \mathcal{A}^N \to \mathcal{A}^N\} = \mathcal{E}$.

If E is a (complex) vector bundle over a compact manifold M of dimension n, there exists a finite cover $\{U_i \ , \ i = 1, \cdots, n\}$ of M such that $E_{|U_i}$ is trivial [82]. Thus, the integer N which determines the rank of the trivial bundle, from whose sections one projects onto the sections of the bundle $E \to M$, is determined by the equality $N = kn$ where k is the rank of the bundle $E \to M$ and n is the dimension of M.

4.3 Hermitian Structures over Projective Modules

Suppose the vector bundle $E \to M$ is also endowed with a Hermitian structure. Then, the Hermitian inner product $\langle \cdot, \cdot \rangle_m$ on each fiber E_m of the bundle gives a $C^\infty(M)$-valued sesquilinear map on the module $\Gamma(E,M)$ of smooth sections,

$$\langle \cdot, \cdot \rangle \; : \; \mathcal{E} \times \mathcal{E} \to C^\infty(M) \; ,$$
$$\langle \eta_1, \eta_2 \rangle (m) =: \langle \eta_1(m), \eta_2(m) \rangle_m \; , \quad \forall \; \eta_1, \eta_2 \in \Gamma(E,M) \; . \quad (4.17)$$

For any $\eta_1, \eta_2 \in \Gamma(E,M)$ and $a, b \in C^\infty(M)$, the map (4.17) is easily seen to satisfy the following properties,

$$\langle \eta_1 a, \eta_2 b \rangle = a^* \langle \eta_1, \eta_2 \rangle b \; , \tag{4.18}$$
$$\langle \eta_1, \eta_2 \rangle^* = \langle \eta_2, \eta_1 \rangle \; , \tag{4.19}$$
$$\langle \eta, \eta \rangle \geq 0 \; , \quad \langle \eta, \eta \rangle = 0 \; \Leftrightarrow \; \eta = 0 \; . \tag{4.20}$$

Suppose now that we have a (finite projective right) module $\mathcal{E}$ over an algebra $\mathcal{A}$ with involution *. Then, equations (4.19)-(4.20) are just the definition of a *Hermitian structure* over $\mathcal{E}$, a module being called *Hermitian* if it admits a Hermitian structure. We recall that an element $a \in \mathcal{A}$ is said to be positive if it can be written in the form $a = b^*b$ for some $b \in \mathcal{A}$.

A condition for non degeneracy of a Hermitian structure is expressed in terms of the *dual module*

$$\mathcal{E}' = \{ \phi : \mathcal{E} \to \mathcal{A} \; | \; \phi(\eta a) = \phi(\eta)a \; , \quad \eta \in \mathcal{E}, a \in \mathcal{A} \} \; , \tag{4.21}$$

to which, $\mathcal{A}$ being a *-algebra, we give a *right* $\mathcal{A}$-module structure as follows,

$$\mathcal{E}' \times \mathcal{A} \ni (\phi, a) \mapsto \phi \cdot a =: a^* \cdot \phi \in \mathcal{E}' \; , \quad (\phi \cdot a)(\eta) =: a^*(\phi(\eta)) \; , \tag{4.22}$$

for $a \in \mathcal{A}$, $\phi \in \mathcal{E}'$, $\eta \in \mathcal{E}$. Notice that the previous structure is different from the *left* structure we have defined on $\mathcal{E}'$ in (4.7) (we should indeed use distinct notations, but in what follows we shall use only this right structure over $\mathcal{E}'$!). A condition for non degeneracy could equivalently be defined in terms of the left structure on $\mathcal{E}'$.

We have the following definition.

Definition 4.3.1. *The Hermitian structure* $\langle \cdot, \cdot \rangle$ *on the (right, finite projective)* $\mathcal{A}$-*module* $\mathcal{E}$ *is called* non degenerate *if the map*

$$\mathcal{E} \to \mathcal{E}' \; , \quad \eta \mapsto \langle \eta, \cdot \rangle \; , \tag{4.23}$$

is an isomorphism.

On the free module $\mathcal{A}^N$ there is a canonical Hermitian structure given by

$$\langle \eta, \xi \rangle = \sum_{j=1}^{N} \eta_j^* \xi_j \; , \tag{4.24}$$

where $\eta = (\eta_1, \cdots, \eta_N)$ and $\xi = (\xi_1, \cdots, \xi_N)$ are any two elements of $\mathcal{A}^N$. Under suitable regularity conditions on the algebra $\mathcal{A}$ all Hermitian structures on a given finite projective module $\mathcal{E}$ over $\mathcal{A}$ are isomorphic to each other and are obtained from the canonical structure (4.24) on $\mathcal{A}^N$ by restriction. We refer to [32] for additional considerations and details on this point. Moreover, if $\mathcal{E} = p\mathcal{A}^N$, then p is self-adjoint.[3] Indeed we have the following proposition

Proposition 4.3.1. *Hermitian finite projective modules are of the form $p\mathcal{A}^N$ with p a self-adjoint idempotent, $p^* = p$, the $*$ operation being the composition of the $*$ operation in the algebra $\mathcal{A}$ with the usual matrix transposition.*

Proof. With respect to the canonical structure (4.24), one easily finds that $\langle p^*\xi, \eta \rangle = \langle \xi, p\eta \rangle$ for any matrix $p \in \mathbb{M}_N(\mathcal{A})$. Now suppose that p is an idempotent and consider the module $\mathcal{E} = p\mathcal{A}^N$. The orthogonal space

$$\mathcal{E}^\perp =: \{ u \in \mathcal{A}^N \mid \langle u, \eta \rangle = 0 \, , \, \forall \, \eta \in \mathcal{E} \} \qquad (4.25)$$

is again a right $\mathcal{A}$-module since $\langle ua, \eta \rangle = a^* \langle u, \eta \rangle$. If $u \in \mathcal{A}^N$ and $\eta \in \mathcal{E}$, then $\langle (1 - p^*)u, \eta \rangle = \langle u, (1 - p)\eta \rangle = 0$ which states that $\mathcal{E}^\perp = (1 - p^*)\mathcal{A}^N$. On the other hand, since $\mathcal{A}^N = p\mathcal{A}^N \oplus (1 - p)\mathcal{A}^N$, the pairing $\langle \cdot, \cdot \rangle$ on $\mathcal{A}^N$ gives a Hermitian structure on $\mathcal{E} = p\mathcal{A}^N$ if and only if this is an orthogonal direct sum, that is, if and only if $(1 - p^*) = (1 - p)$ or $p = p^*$.

4.4 The Algebra of Endomorphisms of a Module

Suppose we are given a Hermitian finite projective $\mathcal{A}$-module $\mathcal{E} = p\mathcal{A}^N$. The *algebra of endomorphisms* of $\mathcal{E}$ is defined by

$$End_\mathcal{A}(\mathcal{E}) = \{ \phi : \mathcal{E} \to \mathcal{E} \mid \phi(\eta a) = \phi(\eta)a \, , \, \eta \in \mathcal{E} \, , \, a \in \mathcal{A} \} \, . \qquad (4.26)$$

It is clearly an algebra under composition. It also admits a natural involution $* : End_\mathcal{A}(\mathcal{E}) \to End_\mathcal{A}(\mathcal{E})$ determined by[4]

$$\langle T^*\eta, \xi \rangle =: \langle \eta, T\xi \rangle \, , \quad \forall \, T \in End_\mathcal{A}(\mathcal{E}) \, , \, \eta, \xi \in \mathcal{E} \, . \qquad (4.27)$$

With this involution, there is an isomorphism

$$End_\mathcal{A}(\mathcal{E}) \simeq p\mathbb{M}_N(\mathcal{A})p \, , \qquad (4.28)$$

so that, elements of $End_\mathcal{A}(\mathcal{E})$ are matrices $m \in \mathbb{M}_N(\mathcal{A})$ which commute with the idempotent p, $pm = mp$.

The group $\mathcal{U}(\mathcal{E})$ of *unitary endomorphisms* of $\mathcal{E}$ is the subgroup of $End_\mathcal{A}(\mathcal{E})$ given by

---

[3] Self-adjoint idempotents are also called projectors.

[4] We are being a bit sloppy here. An endomorphism of a module need not admit an adjoint. For a detailed discussion we refer to App. A.3 and in particular to Definition A.3.4. In fact, one considers only endomorphisms admitting an adjoint and $End_\mathcal{A}(\mathcal{E})$ denotes the algebra of all such endomorphisms.

$$\mathcal{U}(\mathcal{E}) = \{u \in End_A(\mathcal{E}) \mid uu^* = u^*u = \mathbb{I}\} . \tag{4.29}$$

In particular, $\mathcal{U}_N(\mathcal{A}) =: \mathcal{U}(\mathcal{A}^N) = \{u \in \mathbb{M}_N(\mathcal{A}) \mid uu^* = u^*u = \mathbb{I}\}$. Also, there is an isomorphism $\mathcal{U}_N(C^\infty(M)) \simeq C^\infty(M, U(N))$, with M a smooth manifold and $U(N)$ the usual N-dimensional unitary group. In general, if $\mathcal{E} = p\mathcal{A}^N$ with $p^* = p$, one finds that $\mathcal{U}(\mathcal{E}) = p\mathcal{U}(\mathcal{A}^N)p$.

4.5 More Bimodules of Various Kinds

As alluded to before, there are situations in which one needs more than right (or left) modules. We mention in particular the study of 'linear connections', namely connections on the module of one-forms. As we shall see, the latter carries a natural bimodule structure.

Here we briefly describe two relevant constructions which have been proposed in [53]

Definition 4.5.1. *Let $\mathcal{E}$ be a bimodule over the algebra $\mathcal{A}$, with $Z(\mathcal{A})$ denoting the center of the latter. Then, $\mathcal{E}$ is called a* central *bimodule if it happens that*

$$z\eta = \eta z , \quad \forall z \in Z(\mathcal{A}) , \eta \in \mathcal{E} . \tag{4.30}$$

The previous definition just says that the inherited structure of a bimodule over the center $Z(\mathcal{A})$ is *induced* by a structure of $Z(\mathcal{A})$-module. Indeed, if $\mathcal{F}$ is a right (say) module over a *commutative* algebra $\mathcal{C}$, one can induce a $\mathcal{C}$-bimodule structure over $\mathcal{F}$ by defining a left action of $\mathcal{C}$ simply by

$$c\eta =: \eta c , \quad \forall c \in \mathcal{C} , \eta \in \mathcal{F} . \tag{4.31}$$

The commutativity of $\mathcal{C}$ implies that the requirement for a left structure is met. Thus, a central bimodule over a commutative algebra, is just a module with the induced bimodule structure. The category of central bimodules over an algebra $\mathcal{A}$ is stable under the operation of taking tensor products over $Z(\mathcal{A})$ and a fortiori also over $\mathcal{A}$.

Definition 4.5.2. *Let $\mathcal{E}$ be a bimodule over the algebra $\mathcal{A}$. Then, $\mathcal{E}$ is called a* diagonal *bimodule if $\mathcal{E}$ is isomorphic (as a bimodule) to a sub-bimodule of $\mathcal{A}^I$, with $I = I(\mathcal{E})$ any set and $\mathcal{A}$ is equipped with its canonical bimodule structure.*

A diagonal bimodule is central but the converse need not be true. A motivation for the previous definition is that if $\mathcal{A}$ is commutative, then a diagonal bimodule is an $\mathcal{A}$-module (with the induced bimodule structure) such that the canonical map from $\mathcal{E}$ to its bi-dual $\mathcal{E}''$ is injective. The category of diagonal bimodules over an algebra $\mathcal{A}$ is stable under the operation of taking tensor products over $\mathcal{A}$; indeed, if $\mathcal{E} \subset \mathcal{A}^I$ and $\mathcal{F} \subset \mathcal{A}^J$, then $\mathcal{E} \otimes_A \mathcal{F} \subset \mathcal{A}^{I \times J}$.

Another situation where one needs to go beyond 'bare' modules is when analyzing 'real sections of a vector bundle'. A noncommutative analogue of complexified vector bundles is provided by *-bimodules [54]

Definition 4.5.3. *Let $\mathcal{E}$ be a bimodule over the *-algebra $\mathcal{A}$. Then, $\mathcal{E}$ is a *-bimodule if it has an antilinear involution $\mathcal{E} \ni \eta \mapsto \eta^* \in \mathcal{E}$, such that*

$$(a\eta b)^* = b^* \eta^* a^* , \quad a, b \in \mathcal{A} , \ \eta \in \mathcal{E} . \tag{4.32}$$

Given a *-bimodule $\mathcal{E}$, the analogues of real sections are the *-invariant elements of $\mathcal{E}$,

$$\mathcal{E}_r =: \{\eta \in \mathcal{E} \mid \eta^* = \eta\} . \tag{4.33}$$

We refer to [49, 53, 54] for additional details and further discussions.

5. A Few Elements of K-Theory

We have seen in the previous Section that the algebraic substitutes for bundles (of a particular kind, at least) are projective modules of finite type over an algebra $\mathcal{A}$. The (algebraic) K-theory of $\mathcal{A}$ is the natural framework for the analogue of bundle invariants. Indeed, both the notions of isomorphism and of stable isomorphism have a meaning in the context of finite projective (right) modules. The group $K_0(\mathcal{A})$ will be the group of (stable) isomorphism classes of such modules.

In this Section we shall review some of the fundamentals of the K-theory of C^*-algebras while referring to [131, 12] for more details. In particular, we shall have AF algebras in mind.

5.1 The Group K_0

Given a unital C^*-algebra $\mathcal{A}$ we denote by $\mathbb{M}_N(\mathcal{A}) \simeq \mathbb{M}_N(\mathbb{C}) \otimes_{\mathbb{C}} \mathcal{A}$ the C^*-algebra of $N \times N$ matrices with entries in $\mathcal{A}$. Two projectors $p, q \in \mathbb{M}_N(\mathcal{A})$ are said to be *equivalent* (in the sense of Murray-von Neumann) if there exists a matrix (a partial isometry[1]) $u \in \mathbb{M}_N(\mathcal{A})$ such that $p = u^*u$ and $q = uu^*$. In order to be able to 'add' equivalence classes of projectors, one considers all finite matrix algebras over $\mathcal{A}$ at the same time by considering $\mathbb{M}_\infty(\mathcal{A})$ which is the non complete *-algebra obtained as the inductive limit of finite matrices[2],

$$\mathbb{M}_\infty(\mathcal{A}) = \bigcup_{n=1}^{\infty} \mathbb{M}_n(\mathcal{A}) \, ,$$

$$\phi : \mathbb{M}_n(\mathcal{A}) \to \mathbb{M}_{n+1}(\mathcal{A}) \, , \quad a \mapsto \phi(a) = \left\{ \begin{array}{cc} a & 0 \\ 0 & 0 \end{array} \right\} . \tag{5.1}$$

Now, two projectors $p, q \in \mathbb{M}_\infty(\mathcal{A})$ are said to be equivalent, $p \sim q$, when there exists a $u \in \mathbb{M}_\infty(\mathcal{A})$ such that $p = u^*u$ and $q = uu^*$. The set $V(\mathcal{A})$

---

[1] An element u in a *-algebra $\mathcal{B}$ is called a *partial isometry* if u^*u is a projector (called the support projector). Then automatically uu^* is a projector [131] (called the range projector). If $\mathcal{B}$ is unital and $u^*u = \mathbb{I}$, then u is called an *isometry*.

[2] The completion of $\mathbb{M}_\infty(\mathcal{A})$ is $\mathcal{K} \otimes \mathcal{A}$, with $\mathcal{K}$ the algebra of compact operators on the Hilbert space l_2. The algebra $\mathcal{K} \otimes \mathcal{A}$ is also called the stabilization of $\mathcal{A}$.

of equivalence classes $[\,\cdot\,]$ is made into an abelian semigroup by defining an *addition*,

$$[p] + [q] =: [\left\{ \begin{array}{cc} p & 0 \\ 0 & q \end{array} \right\}] , \quad \forall\, [p], [q] \in V(\mathcal{A}). \tag{5.2}$$

The additive identity is just $0 =: [0]$.

The group $K_0(\mathcal{A})$ is the universal canonical group (also called the *enveloping* or *Grothendieck group*) associated with the abelian semigroup $V(\mathcal{A})$. It may be defined as a collection of equivalence classes,

$$\begin{aligned} & K_0(\mathcal{A}) =: V(\mathcal{A}) \times V(\mathcal{A})/ \sim , \\ & ([p], [q]) \sim ([p'], [q']) \\ & \Leftrightarrow \; \exists\, [r] \in V(\mathcal{A}) \;\; \text{s.t.} \;\; [p] + [q'] + [r] = [p'] + [q] + [r] . \end{aligned} \tag{5.3}$$

It is straightforward to check reflexivity, symmetry and transitivity, the extra $[r]$ in (5.3) being inserted just to get the latter property, so that. Thus $\sim$ is an equivalence relation. The presence of the extra $[r]$ is the reason why one is only classifying *stable* classes.

The addition in $K_0(\mathcal{A})$ is defined by

$$[([p], [q])] + [([p'], [q'])] =: [([p] + [p'], [q] + [q'])] , \tag{5.4}$$

for any $[([p], [q])], [([p'], [q'])] \in K_0(\mathcal{A})$, and does not depend on the representatives. As for the neutral element, it is given by the class

$$0 = [([p], [p])] , \tag{5.5}$$

for any $[p] \in V(\mathcal{A})$. Indeed, all such elements are equivalent. Finally, the inverse $-[([p], [q])]$ of the class $[([p], [q])]$ is given by the class

$$- [([p], [q])] =: [([q], [p])] , \tag{5.6}$$

since,

$$[([p], [q])] + (-[([p], [q])]) = [([p], [q])] + ([([q], [p])]) = [([p] + [q], [p] + [q])] = 0 . \tag{5.7}$$

From all that has been previously said, it is clear that it is useful to think of the class $[([p], [q])] \in K_0(\mathcal{A})$ as a *formal difference* $[p] - [q]$.

There is a natural homomorphism

$$\kappa_{\mathcal{A}} : V(\mathcal{A}) \to K_0(\mathcal{A}) , \quad \kappa_{\mathcal{A}}([p]) =: ([p], [0]) = [p] - [0] . \tag{5.8}$$

However, this map is injective if and only if the addition in $V(\mathcal{A})$ has cancellations, namely if and only if $[p] + [r] = [q] + [r] \Rightarrow [p] = [q]$. Independently of the fact that $V(\mathcal{A})$ has cancellations, any $\kappa_{\mathcal{A}}([p]), [p] \in V(\mathcal{A})$, has an inverse in $K_0(\mathcal{A})$ and any element of the latter group can be written as a difference $\kappa_{\mathcal{A}}([p]) - \kappa_{\mathcal{A}}([q])$, with $[p], [q] \in V(\mathcal{A})$.

While for a generic $\mathcal{A}$, the semigroup $V(\mathcal{A})$ has no cancellations, for AF algebras this happens to be the case. By defining

$$K_{0+}(\mathcal{A}) =: \kappa_\mathcal{A}(V(\mathcal{A})) , \tag{5.9}$$

the couple $(K_0(\mathcal{A}), K_{0+}(\mathcal{A}))$ becomes, for an AF algebra $\mathcal{A}$, an *ordered group* with $K_{0+}(\mathcal{A})$ the *positive cone*, namely one has that

$$K_{0+}(\mathcal{A}) \ni 0 ,$$
$$K_{0+}(\mathcal{A}) - K_{0+}(\mathcal{A}) = K_0(\mathcal{A}) ,$$
$$K_{0+}(\mathcal{A}) \cap (-K_{0+}(\mathcal{A})) = 0 . \tag{5.10}$$

For a generic algebra the last property is not true and, as a consequence, the couple $(K_0(\mathcal{A}), K_{0+}(\mathcal{A}))$ is not an ordered group.

Example 5.1.1. The group $K_0(\mathcal{A})$ for $\mathcal{A} = \mathbb{C}$, $\mathcal{A} = \mathbb{M}_k(\mathbb{C})$, $k \in \mathbb{N}$ and $\mathcal{A} = \mathbb{M}_k(\mathbb{C}) \oplus \mathbb{M}_{k'}(\mathbb{C})$, $k, k' \in \mathbb{N}$.
If $\mathcal{A} = \mathbb{C}$, any element in $V(\mathcal{A})$ is a class of equivalent projectors in some $\mathbb{M}_n(\mathbb{C})$. Now, projectors in $\mathbb{M}_n(\mathbb{C})$ are equivalent precisely when their ranges, which are subspaces of $\mathbb{C}^n$, have the same dimension. Therefore we can make the identification

$$V(\mathbb{C}) \simeq \mathbb{N} , \tag{5.11}$$

with $\mathbb{N} = \{0, 1, 2, \cdots\}$ the semigroup of natural numbers.

As $\mathbb{M}_n(\mathbb{M}_k(\mathbb{C})) \simeq \mathbb{M}_{nk}(\mathbb{C})$, the same argument gives

$$V(\mathbb{M}_k(\mathbb{C})) \simeq \mathbb{N} . \tag{5.12}$$

The canonical group associated with the semigroup $\mathbb{N}$ is just the group $\mathbb{Z}$ of integers, and we have

$$K_0(\mathbb{C}) = \mathbb{Z} , \qquad K_{0+}(\mathbb{C}) = \mathbb{N} ,$$
$$K_0(\mathbb{M}_k(\mathbb{C})) = \mathbb{Z} , \quad K_{0+}(\mathbb{M}_k(\mathbb{C})) = \mathbb{N} , \quad \forall k \in \mathbb{N} . \tag{5.13}$$

For $\mathcal{A} = \mathbb{M}_k(\mathbb{C}) \oplus \mathbb{M}_{k'}(\mathbb{C})$, the same argument for each of the two terms in the direct sum will give

$$K_0(\mathbb{M}_k(\mathbb{C}) \oplus \mathbb{M}_{k'}(\mathbb{C})) = \mathbb{Z} \oplus \mathbb{Z} , \tag{5.14}$$
$$K_{0+}(\mathbb{M}_k(\mathbb{C}) \oplus \mathbb{M}_{k'}(\mathbb{C})) = \mathbb{N} \oplus \mathbb{N} , \quad \forall k, k' \in \mathbb{N} . \tag{5.15}$$

In general, the group K_0 has a few interesting properties, notably universality.

Proposition 5.1.1. *Let G be an abelian group and $\Phi : V(\mathcal{A}) \to G$ be a homomorphism of semigroups such that $\Phi(V(\mathcal{A}))$ is invertible in G. Then, Φ extends uniquely to a homomorphism $\Psi : K_0(\mathcal{A}) \to G$,*

$$
\begin{array}{ccc}
\Phi: & V(\mathcal{A}) & \longrightarrow & G \\
& & & \\
& \kappa_\mathcal{A} \downarrow & \uparrow \Psi & , \quad \Psi \circ \kappa_\mathcal{A} = \Phi . \\
& & & \\
id: & K_0(\mathcal{A}) & \longleftrightarrow & K_0(\mathcal{A})
\end{array}
\tag{5.16}
$$

Proof. First uniqueness.

If $\Psi_1, \Psi_2 : K_0(\mathcal{A}) \to G$ both extend Φ, then $\Psi_1([([p], [q])]) = \Psi_1([p] - [q]) = \Psi_1(\kappa_{\mathcal{A}}([p])) - \Psi_1(\kappa_{\mathcal{A}}([q])) = \Phi([p]) - \Phi([q]) = \Psi_2([([p], [q])])$, which proves uniqueness.

Then existence.

Define $\Psi : K_0(\mathcal{A}) \to G$ by $\Psi([([p], [q])]) = \Phi([p]) - \Phi([q])$. This map is well defined because $\Phi([q])$ has an inverse in G and because $([p], [q]) \sim ([p'], [q']) \Leftrightarrow \exists [r] \in V(\mathcal{A})$ such that $[p] + [q'] + [r] = [p'] + [q] + [r]$, and this in turn, implies $\Psi([([p], [q])]) = \Psi([([p'], [q'])])$. Finally, Ψ is a homomorphism and $\Psi(\kappa_{\mathcal{A}}([p]) = \Psi([([p], [0])]) = \Phi([p])$, i.e. $\Psi \circ \kappa_{\mathcal{A}} = \Phi$.

The group K_0 is well behaved with respect to homomorphisms.[3]

Proposition 5.1.2. *If $\alpha : \mathcal{A} \to \mathcal{B}$ is a homomorphism of C^*-algebras, then the induced map*

$$\alpha_* : V(\mathcal{A}) \to V(\mathcal{B}) , \quad \alpha_*([a_{ij}]) =: [\alpha(a_{ij})] , \tag{5.17}$$

is a well defined homomorphism of semigroups. Moreover, from universality, α_* extends to a group homomorphism (denoted by the same symbol)

$$\alpha_* = K_0(\mathcal{A}) \to K_0(\mathcal{B}) . \tag{5.18}$$

Proof. If the matrix $(a_{ij}) \in \mathbb{M}_\infty(\mathcal{A})$ is a projector, the matrix $\alpha(a_{ij})$ will clearly be a projector in $\mathbb{M}_\infty(\mathcal{B})$ since α is a homomorphism. Furthermore, if (a_{ij}) is equivalent to (b_{ij}), then, since α is multiplicative and *-preserving, $\alpha(a_{ij})$ will be equivalent to $\alpha(b_{ij})$. Thus $\alpha_* : V(\mathcal{A}) \to V(\mathcal{B})$ is well defined and clearly a homomorphism. The last statement follows from Proposition 5.1.1 with the identification $\Phi \equiv \kappa_{\mathcal{B}} \circ \alpha_* : V(\mathcal{A}) \to K_0(\mathcal{B})$ so as to get for Ψ the map $\Psi \equiv \alpha_* : K_0(\mathcal{A}) \to K_0(\mathcal{B})$.

The group K_0 is also well behaved with respect to the process of taking inductive limits of C^*-algebras, as stated in the following proposition which is proven in [131] and which is crucial for the calculation of the group K_0 of AF algebras.

Proposition 5.1.3. *If the C^*-algebra $\mathcal{A}$ is the inductive limit of an inductive system $\{\mathcal{A}_i, \Phi_{ij}\}_{i,j \in \mathbb{N}}$ of C^*-algebras[4], then $\{K_0(\mathcal{A}_i), \Phi_{ij*}\}_{i,j \in \mathbb{N}}$ is an inductive system of groups and one can exchange the limits,*

$$K_0(\mathcal{A}) = K_0(\lim_{\to} \mathcal{A}_i) = \lim_{\to} K_0(\mathcal{A}_i) . \tag{5.19}$$

Moreover, if $\mathcal{A}$ is an AF algebra, then $K_0(\mathcal{A})$ is an ordered group with the positive cone given by the limit of an inductive system of semigroups

$$K_{0+}(\mathcal{A}) = K_{0+}(\lim_{\to} \mathcal{A}_i) = \lim_{\to} K_{0+}(\mathcal{A}_i) . \tag{5.20}$$

---

[3] In a more sophisticated parlance, K_0 is a covariant functor from the category of C^*-algebras to the category of abelian groups.

[4] In fact, one could substitute $\mathbb{N}$ with any directed set Λ.

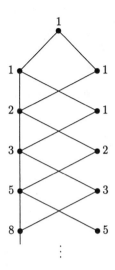

Fig. 5.1. The Bratteli diagram for the algebra $\mathcal{A}_{PT}$ of the Penrose tiling

One has that as sets,

$$K_0(\mathcal{A}) = \{(k_n)_{n\in\mathbb{N}} , k_n \in K_0(\mathcal{A}_n) \mid \exists N_0 , k_{n+1} = T_n(k_n) , n > N_0\} , \tag{5.21}$$

$$K_{0+}(\mathcal{A}) = \{(k_n)_{n\in\mathbb{N}} , k_n \in K_{0+}(\mathcal{A}_n) \mid \exists N_0 , k_{n+1} = T_n(k_n) , n > N_0\} , \tag{5.22}$$

while the (abelian) group/semigroup structure is inherited pointwise from the addition in the groups/semigroups in the sequences (5.21), (5.22) respectively.

5.2 The K-Theory of the Penrose Tiling

The algebra $\mathcal{A}_{PT}$ of the Penrose Tiling is an AF algebra which is quite far from being postliminal, since there are an infinite number of non equivalent irreducible representations which are faithful. These have the same kernel, namely $\{0\}$, which is the only primitive ideal (the algebra $\mathcal{A}_{PT}$ is indeed simple). The construction of its K-theory is rather straightforward and quite illuminating. The corresponding Bratteli diagram is shown in Fig. 5.1 [32]. From Props. (3.4.3) and (3.4.4) it is clear that $\{0\}$ is the only primitive ideal.

At each level, the algebra is given by

$$\mathcal{A}_n = \mathbb{M}_{d_n}(\mathbb{C}) \oplus \mathbb{M}_{d_n'}(\mathbb{C}) , \quad n \geq 1 , \tag{5.23}$$

with inclusion

$$I_n : A_n \hookrightarrow A_{n+1} , \quad \left\{ \begin{array}{cc} A & 0 \\ 0 & B \end{array} \right\} \mapsto \left\{ \begin{array}{ccc} A & 0 & 0 \\ 0 & B & 0 \\ 0 & 0 & A \end{array} \right\} , \quad (5.24)$$

for any $A \in \mathbb{M}_{d_n} (\mathbb{C})$, $B \in \mathbb{M}_{d'_n} (\mathbb{C})$. This gives for the dimensions the following recursive relations,

$$\begin{array}{ll} d_{n+1} = d_n + d'_n , \\ d'_{n+1} = d_n , \end{array} \quad n \geq 1 , \; d_1 = d'_1 = 1 . \quad (5.25)$$

From what we said in Example 5.1.1, after the second level, the K-groups are given by

$$K_0(A_n) = \mathbb{Z} \oplus \mathbb{Z} , \quad K_{0+}(A_n) = \mathbb{N} \oplus \mathbb{N} , \; n \geq 1. \quad (5.26)$$

The group $(K_0(A), K_{0+}(A))$ is obtained, by Proposition 5.1.3, as the inductive limit of the sequence of groups/semigroups

$$K_0(A_1) \hookrightarrow K_0(A_2) \hookrightarrow K_0(A_3) \hookrightarrow \cdots \quad (5.27)$$
$$K_{0+}(A_1) \hookrightarrow K_{0+}(A_2) \hookrightarrow K_{0+}(A_3) \hookrightarrow \cdots \quad (5.28)$$

The inclusions

$$T_n : K_0(A_n) \hookrightarrow K_0(A_{n+1}) , \quad T_n : K_{0+}(A_n) \hookrightarrow K_{0+}(A_{n+1}) , \quad (5.29)$$

are easily obtained from the inclusions I_n in (5.24), being indeed the corresponding induced maps as in (5.18) $T_n = I_{n*}$. To construct the maps T_n we need the following proposition, the first part of which is just Proposition 3.4.1 which we repeat here for clarity.

Proposition 5.2.1. *Let A and B be the direct sum of two matrix algebras,*

$$A = \mathbb{M}_{p_1} (\mathbb{C}) \oplus \mathbb{M}_{p_2} (\mathbb{C}) , \quad B = \mathbb{M}_{q_1} (\mathbb{C}) \oplus \mathbb{M}_{q_2} (\mathbb{C}) . \quad (5.30)$$

Then, any (unital) homomorphism $\alpha : A \to B$ can be written as the direct sum of the representations $\alpha_j : A \to \mathbb{M}_{q_j} (\mathbb{C}) \simeq B(\mathbb{C}^{q_j})$, $j = 1, 2$. If π_{ji} is the unique irreducible representation of $\mathbb{M}_{p_i} (\mathbb{C})$ in $B(\mathbb{C}^{q_j})$, then α_j breaks into a direct sum of the π_{ji}. Furthermore, let N_{ji} be the non-negative integers denoting the multiplicity of π_{ji} in this sum. Then the induced homomorphism, $\alpha_* = K_0(A) \to K_0(B)$, is given by the 2×2 matrix (N_{ij}).

Proof. For the first part just refer to Proposition 3.4.1.
Furthermore, given a rank k projector in $\mathbb{M}_{p_i} (\mathbb{C})$, the representation α_j sends it to a rank $N_{ji}k$ projector in $\mathbb{M}_{q_j} (\mathbb{C})$. This proves the final statement of the proposition.

For the inclusion (5.2.1), Proposition 5.2.1 gives immediately that the maps (5.29) are both represented by the integer valued matrix

$$T = \left\{ \begin{matrix} 1 & 1 \\ 1 & 0 \end{matrix} \right\} , \qquad (5.31)$$

for any level n. The action of the matrix (5.31) can be represented pictorially as in Fig. 5.2 where the couples (a, b), (a', b') are both in $\mathbb{Z} \oplus \mathbb{Z}$ or $\mathbb{N} \oplus \mathbb{N}$.

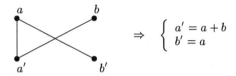

$$\Rightarrow \quad \left\{ \begin{matrix} a' = a + b \\ b' = a \end{matrix} \right.$$

Fig. 5.2. The action of the inclusion T

Finally, we can construct the K_0 group.

Proposition 5.2.2. *The group* $(K_0(\mathcal{A}_{PT}), K_{0+}(\mathcal{A}_{PT}))$ *for the C^*-algebra $\mathcal{A}_{PT}$ of the Penrose tiling is given by*

$$K_0(\mathcal{A}_{PT}) = \mathbb{Z} \oplus \mathbb{Z} , \qquad (5.32)$$

$$K_{0+}(\mathcal{A}_{PT}) = \{ (a, b) \in \mathbb{Z} \oplus \mathbb{Z} : \frac{1 + \sqrt{5}}{2} a + b \geq 0 \} . \qquad (5.33)$$

Proof. The result (5.32) follows immediately from the fact that the matrix T in (5.31) is invertible over the integers, its inverse being

$$T^{-1} = \left\{ \begin{matrix} 0 & 1 \\ 1 & -1 \end{matrix} \right\} . \qquad (5.34)$$

Now, from the definition of inductive limit we have that,

$$K_0(\mathcal{A}_{PT}) = \{ (k_n)_{n \in \mathbb{N}} , k_n \in K_0(\mathcal{A}_n) \mid \exists N_0 , k_{n+1} = T(k_n) , n > N_0 \}. \qquad (5.35)$$

Since T is a bijection, for any $k_{n+1} \in K_0(\mathcal{A}_{n+1})$, there exists a unique $k_n \in K_0(\mathcal{A}_n)$ such that $k_{n+1} = T k_n$. Thus, $K_0(\mathcal{A}_{PT}) = K_0(\mathcal{A}_n) = \mathbb{Z} \oplus \mathbb{Z}$.

As for (5.33), since T is *not* invertible over $\mathbb{N}$, $K_{0+}(\mathcal{A}_{PT}) \neq \mathbb{N} \oplus \mathbb{N}$. To construct $K_{0+}(\mathcal{A}_{PT})$, we study the image $T(K_{0+}(\mathcal{A}_n))$ in $K_{0+}(\mathcal{A}_{n+1})$. It is easily found to be

$$\begin{aligned} T(K_{0+}(\mathcal{A}_n)) &= \{ (a_{n+1}, b_{n+1}) \in \mathbb{N} \oplus \mathbb{N} : a_{n+1} \geq b_{n+1} \} \\ &\neq K_{0+}(\mathcal{A}_{n+1}) . \end{aligned} \qquad (5.36)$$

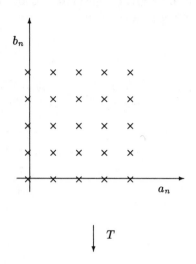

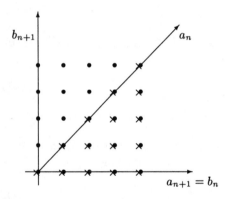

Fig. 5.3. The image of $\mathbb{N} \oplus \mathbb{N}$ under T

Now, T being injective, $T(K_{0+}(\mathcal{A}_n)) = T(\mathbb{N} \oplus \mathbb{N}) \simeq \mathbb{N} \oplus \mathbb{N}$. The inclusion of $T(K_{0+}(\mathcal{A}_n))$ into $K_{0+}(\mathcal{A}_{n+1})$ is shown in Fig. 5.3. By identifying the subset $T(K_{0+}(\mathcal{A}_n)) \subset K_{0+}(\mathcal{A}_{n+1})$ with $K_{0+}(\mathcal{A}_n)$, we can think of $T^{-1}(K_{0+}(\mathcal{A}_{n+1}))$ as a subset of $\mathbb{Z} \oplus \mathbb{Z}$ and of $T^{-1}(K_{0+}(\mathcal{A}_n))$ as the standard positive cone $\mathbb{N} \oplus \mathbb{N}$. The result is shown in Fig. 5.4. The next iteration, namely $T^{-2}(K_{0+}(\mathcal{A}_n))$ is shown in Fig. 5.5. From definition (5.22), by going to the limit we shall have $K_{0+}(\mathcal{A}_{PT}) = \lim_{m \to \infty} T^{-m}(\mathbb{N} \oplus \mathbb{N})$ and the limit will be a subset of $\mathbb{Z} \oplus \mathbb{Z}$ since T is invertible only over $\mathbb{Z}$. The limit can be easily found. From the defining relation

$$F_{m+1} = F_m + F_{m-1}, \; m \geq 1 \,, \tag{5.37}$$

for the Fibonacci numbers, with $F_0 = 0$, $F_1 = 1$, it follows that

$$T^{-m} = (-1)^m \left\{ \begin{array}{cc} F_{m-1} & -F_m \\ -F_m & F_{m+1} \end{array} \right\}. \tag{5.38}$$

Therefore, T^{-m} takes the positive axis $\{(a, 0) : a \geq 0\}$ to a half-line of slope $-F_m/F_{m-1}$, and the positive axis $\{(0, b) : b \geq 0\}$ to a half-line of slope $-F_{m+1}/F_m$. Thus the positive cone $\mathbb{N} \oplus \mathbb{N}$ opens into a fan-shaped wedge which is bordered by these two half-lines. Any integer coordinate point within the wedge comes from an integer coordinate point in the original positive cone. Since $\lim_{m \to \infty} F_{m+1}/F_m = \frac{1+\sqrt{5}}{2}$, the limit cone is just the half-space $\{(a, b) \in \mathbb{Z} \oplus \mathbb{Z} : \frac{1+\sqrt{5}}{2} a + b \geq 0\}$. Every integer coordinate point in it belongs to some intermediate wedge and so it lies in $K_{0+}(\mathcal{A}_{PT})$. The latter is shown in Fig. 5.6.

We refer to [60] for an extensive study of the K-theory of noncommutative lattices and for several examples of K-groups.

5.3 Higher-Order K-Groups

In order to define higher order groups, one needs to introduce the notion of *suspension* of a C^*-algebra $\mathcal{A}$: it is the C^*-algebra

$$S\mathcal{A} =: \mathcal{A} \otimes C_0(\mathbb{R}) \simeq C_0(\mathbb{R} \to \mathcal{A}) , \qquad (5.39)$$

where C_0 indicates continuous functions vanishing at infinity. Also, in the second object, sum and product are defined pointwise, adjoint is the adjoint in $\mathcal{A}$ and the norm is the supremum norm $\|f\|_{S\mathcal{A}} = \sup_{x \in \mathbb{R}} \|f(x)\|_{\mathcal{A}}$.

The K-group of order n of $\mathcal{A}$ is defined to be

$$K_n(\mathcal{A}) =: K_0(S^n \mathcal{A}) , \quad n \in \mathbb{N} . \qquad (5.40)$$

However, the Bott periodicity theorem asserts that all K-groups are isomorphic to either K_0 or K_1, so that there are really only two such groups. There are indeed the following isomorphisms [131]

$$K_{2n}(\mathcal{A}) \simeq K_0(\mathcal{A}) ,$$
$$K_{2n+1}(\mathcal{A}) \simeq K_1(\mathcal{A}) , \quad \forall\, n \in \mathbb{N} . \qquad (5.41)$$

Again, AF algebras show characteristic features. Indeed, for them K_1 vanishes identically.

While K-theory provides analogues of topological invariants for algebras, cyclic cohomology provides analogues of differential geometric invariants. K-theory and cohomology are connected by the noncommutative Chern character in a beautiful generalization of the usual (commutative) situation [32]. We regret that all this goes beyond the scope of the present notes.

As mentioned in Sect. 3.4.2, K-theory has been proven [58] to be a complete invariant which distinguishes among AF algebras if one add to the ordered group $(K_0(\mathcal{A}), K_{0+}(\mathcal{A}))$ the notion of *scale*, the latter being defined for any C^*-algebra $\mathcal{A}$ as

$$\Sigma \mathcal{A} =: \{[p] , \; p \text{ a projector in } \mathcal{A}\} . \qquad (5.42)$$

AF algebras are completely determined, up to isomorphism, by their *scaled ordered* groups, namely by the triples (K_0, K_{0+}, Σ). The key to this is the fact that scale preserving isomorphisms between the ordered groups (K_0, K_{0+}, Σ) of two AF algebras are nothing but K-theoretically induced maps (5.18) of isomorphisms between the AF algebras themselves.

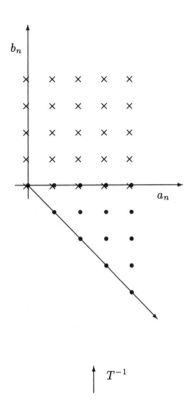

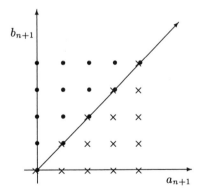

Fig. 5.4. The image of $\mathbb{N} \oplus \mathbb{N}$ under T^{-1}

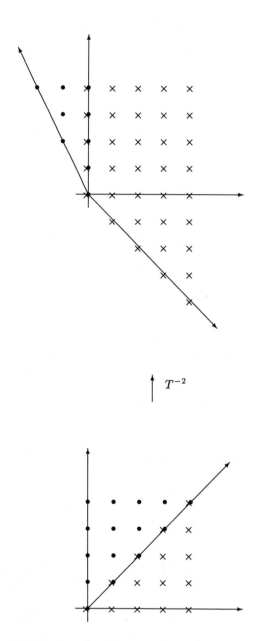

Fig. 5.5. The image of $\mathbb{N} \oplus \mathbb{N}$ under T^{-2}

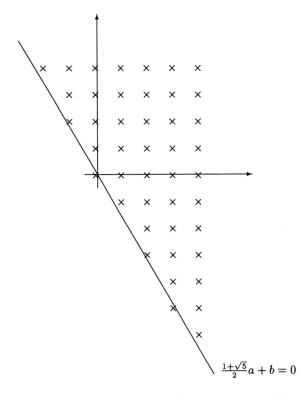

Fig. 5.6. The semigroup $K_{0+}(\mathcal{A}_{PT})$ for the algebra of the Penrose tiling

6. The Spectral Calculus

In this section we shall introduce the machinery of spectral calculus which is the noncommutative generalization of the usual calculus on a manifold. As we shall see, a crucial rôle is played by the Dixmier trace.

6.1 Infinitesimals

Before we proceed to illustrate Connes' theory of infinitesimals, we need a few additional facts about compact operators which we take from [117, 120] and state as propositions. The algebra of compact operators on the Hilbert space $\mathcal{H}$ will be denoted by $\mathcal{K}(\mathcal{H})$ while $\mathcal{B}(\mathcal{H})$ will be the algebra of bounded operators.

Proposition 6.1.1. *Let T be a compact operator on $\mathcal{H}$. Then, its spectrum $\sigma(T)$ is a discrete set having no limit points except perhaps $\lambda = 0$. Furthermore, any nonzero $\lambda \in \sigma(T)$ is an eigenvalue of finite multiplicity.*

Notice that a generic compact operator needs not admit any nonzero eigenvalue.

Proposition 6.1.2. *Let T be a self-adjoint compact operator on $\mathcal{H}$. Then, there is a complete orthonormal basis, $\{\phi_n\}_{n\in\mathbb{N}}$, for $\mathcal{H}$ of eigenvectors, so that $T\phi_n = \lambda_n\phi_n$ and $\lambda_n \to 0$ as $n \to \infty$.*

Proposition 6.1.3. *Let T be a compact operator on $\mathcal{H}$. Then, it has a uniformly convergent (i.e. convergent in norm) expansion*

$$T = \sum_{n \geq 0} \mu_n(T) |\psi_n\rangle \langle \phi_n| , \qquad (6.1)$$

where, $0 \leq \mu_{j+1} \leq \mu_j$, and $\{\psi_n\}_{n\in\mathbb{N}}, \{\phi_n\}_{n\in\mathbb{N}}$ are (not necessarily complete) orthonormal sets.

In this proposition one writes the polar decomposition $T = U|T|$, with $|T| = \sqrt{T^*T}$. Then, $\{\mu_n(T), \mu_n \to 0$ as $n \to \infty\}$ are the non vanishing eigenvalues of the (compact self-adjoint) operator $|T|$ arranged with repeated multiplicity; $\{\phi_n\}$ are the corresponding eigenvectors and $\psi_n = U\phi_n$. The eigenvalues

$\{\mu_n(T)\}$ are called the *characteristic values* of T. One has that $\mu_0(T) = \|T\|$, the norm of T.

Due to condition (2.38), compact operators are in a sense 'small'; they play the rôle of *infinitesimals*. The size of the infinitesimal $T \in \mathcal{K}(\mathcal{H})$ is governed by the rate of decay of the sequence $\{\mu_n(T)\}$ as $n \to \infty$.

Definition 6.1.1. *For any $\alpha \in \mathbb{R}^+$, the infinitesimals of order α are all $T \in \mathcal{K}(\mathcal{H})$ such that*

$$\mu_n(T) = O(n^{-\alpha}) \text{ , as } n \to \infty \text{ ,}$$
$$i.e. \ \exists \, C < \infty \ : \ \mu_n(T) \le Cn^{-\alpha} \text{ , } \forall \, n \ge 1 \text{ .} \tag{6.2}$$

Given any two compact operators T_1 and T_2, there is a submultiplicative property [120]

$$\mu_{n+m}(T_1 T_2) \le \mu_n(T_1)\mu_n(T_2) \text{ ,} \tag{6.3}$$

which, in turn, implies that the orders of infinitesimals behave well,

$$T_j \text{ of order } \alpha_j \ \Rightarrow \ T_1 T_2 \text{ of order } \le \alpha_1 + \alpha_2 \text{ .} \tag{6.4}$$

Also, infinitesimals of order α form a (not closed) two-sided ideal in $\mathcal{B}(\mathcal{H})$, since for any $T \in \mathcal{K}(\mathcal{H})$ and $B \in \mathcal{B}(\mathcal{H})$, one has that [120],

$$\mu_n(TB) \le \|B\| \, \mu_n(T) \text{ ,}$$
$$\mu_n(BT) \le \|B\| \, \mu_n(T) \text{ .} \tag{6.5}$$

6.2 The Dixmier Trace

As in ordinary differential calculus one looks for an 'integral' which neglects all infinitesimals of order > 1. This is achieved here with the Dixmier trace which is constructed in such a way that

1. Infinitesimals of order 1 are in the domain of the trace.
2. Infinitesimals of order higher than 1 have vanishing trace.

The usual trace is not appropriate. Its domain is the two-sided ideal $\mathcal{L}^1$ of trace class operators. For any $T \in \mathcal{L}^1$, the trace, defined as

$$tr \ T =: \sum_n \langle T\xi_n, \xi_n \rangle \text{ ,} \tag{6.6}$$

is independent of the orthonormal basis $\{\xi_n\}_{n \in \mathbb{N}}$ of $\mathcal{H}$ and is, indeed, the sum of eigenvalues of T. When the latter is positive and compact, one has that

$$tr \ T =: \sum_0^\infty \mu_n(T) \text{ .} \tag{6.7}$$

In general, an infinitesimal of order 1 is not in $\mathcal{L}^1$, since the only control on its characteristic values is that $\mu_n(T) \leq C\frac{1}{n}$, for some positive constant C. Moreover, $\mathcal{L}^1$ contains infinitesimals of order higher than 1. However, for (positive) infinitesimals of order 1, the usual trace (6.7) is at most logarithmically divergent since

$$\sum_0^{N-1} \mu_n(T) \leq C \ln N .$$
(6.8)

The Dixmier trace is just a way to extract the coefficient of the logarithmic divergence. It is somewhat surprising that this coefficient behaves as a trace [44].

We shall indicate with $\mathcal{L}^{(1,\infty)}$ the ideal of compact operators which are infinitesimal of order 1. If $T \in \mathcal{L}^{(1,\infty)}$ is positive, one tries to define a positive functional by taking the limit of the cut-off sums,

$$\lim_{N\to\infty} \frac{1}{\ln N} \sum_0^{N-1} \mu_n(T) .$$
(6.9)

There are two problems with the previous formula: its linearity and its convergence. For any compact operator T, consider the sums,

$$\sigma_N(T) = \sum_0^{N-1} \mu_n(T) , \quad \gamma_N(T) = \frac{\sigma_N(T)}{\ln N} .$$
(6.10)

They satisfy [32],

$$\sigma_N(T_1 + T_2) \leq \sigma_N(T_1) + \sigma_N(T_2) , \quad \forall \; T_1, T_2 ,$$
$$\sigma_{2N}(T_1 + T_2) \geq \sigma_N(T_1) + \sigma_N(T_2) , \quad \forall \; T_1, T_2 > 0 .$$
(6.11)

In turn, for any two positive operators T_1 and T_2,

$$\gamma_N(T_1 + T_2) \leq \gamma_N(T_1) + \gamma_N(T_2) \leq \gamma_{2N}(T_1 + T_2)(1 + \frac{\ln 2}{\ln N}) .$$
(6.12)

From this, we see that linearity would follow from convergence. In general, however, the sequence $\{\gamma_N\}$, although bounded, is not convergent. Notice that, since the eigenvalues $\mu_n(T)$ are unitary invariant, so is the sequence $\{\gamma_N\}$. Therefore, one gets a unitary invariant positive trace on the positive part of $\mathcal{L}^{(1,\infty)}$ for each linear form $\lim_\omega$ on the space $\ell^\infty(\mathbb{N})$ of bounded sequences, satisfying the following conditions:

1. $\lim_\omega \{\gamma_N\} \geq 0$, if $\gamma_N \geq 0$.
2. $\lim_\omega \{\gamma_N\} = \lim\{\gamma_N\}$, if $\{\gamma_N\}$ is convergent, with lim the usual limit.
3. $\lim_\omega \{\gamma_1, \gamma_1, \gamma_2, \gamma_2, \gamma_3, \gamma_3, \cdots\} = \lim_\omega \{\gamma_N\}$.
3'. $\lim_\omega \{\gamma_{2N}\} = \lim_\omega \{\gamma_N\}$.   Scale invariance.

Dixmier proved that there exists an infinity of such scale invariant forms [44, 32]. Associated with any one of them there is a trace, the *Dixmier trace*

$$tr_\omega(T) = \lim{}_\omega \frac{1}{\ln N} \sum_0^{N-1} \mu_n(T) , \quad \forall\, T \geq 0 , \; T \in \mathcal{L}^{(1,\infty)} . \tag{6.13}$$

From (6.12), it also follows that tr_ω is additive on positive operators,

$$tr_\omega(T_1 + T_2) = tr_\omega(T_1) + tr_\omega(T_2) , \quad \forall\, T_1, T_2 \geq 0 , \; T_1, T_2 \in \mathcal{L}^{(1,\infty)} . \tag{6.14}$$

This, together with the fact that $\mathcal{L}^{(1,\infty)}$ is generated by its positive part (see below), implies that tr_ω extends by linearity to the entire $\mathcal{L}^{(1,\infty)}$ with the properties,

1. $tr_\omega(T) \geq 0$ if $T \geq 0$.
2. $tr_\omega(\lambda_1 T_1 + \lambda_2 T_2) = \lambda_1 tr_\omega(T_1) + \lambda_2 tr_\omega(T_2)$.
3. $tr_\omega(BT) = tr_\omega(TB) , \quad \forall\, B \in \mathcal{B}(\mathcal{H})$.
4. $tr_\omega(T) = 0$, if T is of order higher than 1.

Property 3. follows from (6.5). The last property follows from the fact that the space of all infinitesimals of order higher than 1 forms a two-sided ideal whose elements satisfy

$$\mu_n(T) = o(\frac{1}{n}) , \quad i.e. \quad n\mu_n(T) \to 0 , \quad \text{as } n \to \infty . \tag{6.15}$$

As a consequence, the corresponding sequence $\{\gamma_N\}$ is convergent and converges to zero. Therefore, for such operators the Dixmier trace vanishes.

To prove that $\mathcal{L}^{(1,\infty)}$ is generated by its positive part one can use a polar decomposition and the fact that $\mathcal{L}^{(1,\infty)}$ is an ideal. If $T \in \mathcal{L}^{(1,\infty)}$, by considering the self-adjoint and anti self-adjoint parts separately one can suppose that T is self-adjoint. Then, $T = U|T|$ with $|T| = \sqrt{T^2}$ and U is a sign operator, $U^2 = U$; from this $|T| = UT$ and $|T| \in \mathcal{L}^{(1,\infty)}$. Furthermore, one has the decomposition $U = U_+ - U_-$ with $U_\pm = \frac{1}{2}(\mathbb{I} \pm U)$ its spectral projectors (projectors on the eigenspaces with eigenvalue $+1$ and -1 respectively). Therefore, $T = U|T| = U_+|T| - U_-|T| = U_+|T|U_+ - U_-|T|U_-$ is a difference of two positive elements in $\mathcal{L}^{(1,\infty)}$.

In many examples of interest in physics, like Yang-Mills and gravity theories, the sequence $\{\gamma_N\}$ itself converges. In these cases, the limit is given by (6.9) and does not depends on ω.

The following examples are mainly taken from [130].

Example 6.2.1. Powers of the Laplacian on the n-dimensional flat torus T^n. The operator

$$\Delta = -(\frac{\partial^2}{\partial x_1^2} + \cdots + \frac{\partial^2}{\partial x_n^2}) , \tag{6.16}$$

has eigenvalues $||l_j||^2$ where the l_j's are all points of the lattice $\mathbb{Z}^n$ taken with multiplicity one. Thus, $|\Delta|^s$ will have eigenvalues $||l_j||^{2s}$. For the corresponding Dixmier trace, one needs to estimate $(logN)^{-1}\sum_1^N ||l_j||^{2s}$ as $N \to \infty$. Let N_R be the number of lattice points in the ball of radius R centered at the origin of $\mathbb{R}^n$. Then $N_R \sim vol\{x \mid ||x|| \leq R\}$ and $N_{r-dr} - N_r \sim \Omega_n r^{n-1} dr$. Here $\Omega_n = 2\pi^{n/2}/\Gamma(n/2)$ is the area of the unit sphere S^{n-1}. Thus,

$$\sum_{||l|| \leq R} ||l||^{2s} \sim \int_1^\infty r^{2s}(N_{r-dr} - N_r)$$

$$= \Omega_n \int_1^\infty r^{2s+n-1} dr . \tag{6.17}$$

On the other side, $logN_R \sim nlogR$. As $R \to \infty$, we have to distinguish three cases.

For $s > -n/2$,

$$(logN_R)^{-1} \sum_{||l|| \leq R} ||l||^{2s} \to \infty . \tag{6.18}$$

For $s < -n/2$,

$$(logN_R)^{-1} \sum_{||l|| \leq R} ||l||^{2s} \to 0 . \tag{6.19}$$

For $s = -n/2$,

$$(logN_R)^{-1} \sum_{||l|| \leq R} ||l||^{-n} \sim \frac{\Omega_n logR}{nlogR} = \frac{\Omega_n}{n} . \tag{6.20}$$

Therefore, the sequence $\{\gamma_N(|\Delta|^s)\}$ diverges for $s > -n/2$, vanishes for $s < -n/2$ and converges for $s = -n/2$. Thus $\Delta^{-n/2}$ is an infinitesimal of order 1, its trace being given by

$$tr_\omega(\Delta^{-n/2}) = \frac{\Omega_n}{n} = \frac{2\pi^{n/2}}{n\Gamma(n/2)} . \tag{6.21}$$

Example 6.2.2. Powers of the Laplacian on the n-dimensional sphere S^n. The Laplacian operator Δ on S^n has eigenvalues $l(l+n-1)$ with multiplicity

$$m_l = \binom{l+n}{n} - \binom{l+n-2}{n} = \frac{(l+n-1)!}{(n-1)!l!}\frac{(2l+n-1)}{(l+n-1)} , \tag{6.22}$$

where $l \in \mathbb{N}$; in particular $m_0 = 1, m_1 = n+1$. One needs to estimate, as $N \to \infty$, the following sums

$$log\sum_{l=0}^N m_l , \quad \sum_{l=0}^N m_l[l(l+n-1)]^{-n/2} . \tag{6.23}$$

One finds that

$$\sum_{l=0}^{N} m_l = \binom{N+n}{n} + \binom{M+n-1}{n}$$

$$= \frac{1}{n!}(N+n-1)(N+n-2)\cdots(N+1)(2N+n)$$

$$\sim \frac{2N^n}{n!}, \tag{6.24}$$

from which,

$$\log \sum_{l=0}^{N} m_l \sim \log N^n + \log 2 - \log n! \sim n \log N . \tag{6.25}$$

Furthermore,

$$\sum_{l=0}^{N} m_l [l(l+n-1)]^{-n/2} = \frac{1}{(n-1)!} \sum_{l=0}^{N} \frac{(l+n-1)!}{l![l(l+n-1)]^{n/2}} \frac{(2l+n-1)}{(l+n-1)}$$

$$\sim \frac{2}{(n-1)!} \sum_{l=0}^{N} \frac{l^{n-1}}{[l(l+n-1)]^{n/2}}$$

$$\sim \frac{2}{(n-1)!} \sum_{l=0}^{N} \frac{l^{n-1}}{(l+\frac{n-1}{2})^n}$$

$$\sim \frac{2}{(n-1)!} \sum_{l=0}^{N} (l+\frac{n-1}{2})^{-1}$$

$$\sim \frac{2}{(n-1)!} \log N. \tag{6.26}$$

By putting the numerator and the denominator together we finally get,

$$tr_\omega(\Delta^{-n/2}) = \lim_{N\to\infty} \left(\sum_{l=0}^{N} m_l [l(l+n-1)]^{-n/2} / \log \sum_{l=0}^{N} m_l\right)$$

$$= \lim_{N\to\infty} \frac{2\log N/(n-1)!}{n\log N} = \frac{2}{n!} . \tag{6.27}$$

If one replaces the exponent $-n/2$ by a smaller s, the series in (6.26) becomes convergent and the Dixmier trace vanishes. On the other end, if $s > n/2$, this series diverges faster than the one in the denominator and the corresponding quotient diverges.

Example 6.2.3. The inverse of the harmonic oscillator.
The Hamiltonian of the one dimensional harmonic oscillator is given (in 'momentum space') by $H = \frac{1}{2}(\xi^2 + x^2)$. It is well known that on the Hilbert space $L^2(\mathbb{R})$ its eigenvalues are $\mu_n(H) = n + \frac{1}{2}$, $n = 0, 1, \ldots$, while its inverse $H^{-1} = 2(\xi^2 + x^2)^{-1}$ has eigenvalues $\mu_n(H^{-1}) = \frac{2}{2n+1}$. The sequence $\{\gamma_N(H^{-1})\}$ converges and the corresponding Dixmier trace is given by (6.9),

$$tr_\omega(H^{-1}) = \lim_{N\to\infty} \frac{1}{\ln N} \sum_0^{N-1} \mu_n(H^{-1}) = \lim_{N\to\infty} \frac{1}{\ln N} \sum_0^{N-1} \frac{2}{2n+1} = 1 \;. \quad (6.28)$$

6.3 Wodzicki Residue and Connes' Trace Theorem

The Wodzicki(-Adler-Manin-Guillemin) residue [134] is the unique trace on the algebra of pseudodifferential operators of any order which, on operators of order at most $-n$ coincides with the corresponding Dixmier trace. Pseudo-differential operators are briefly described in App. A.6. In this section we shall introduce the residue and the theorem by Connes [31] which establishes its connection with the Dixmier trace.

Definition 6.3.1. *Let M be an n-dimensional compact Riemannian manifold. Let T be a pseudodifferential operator of order $-n$ acting on sections of a complex vector bundle $E \to M$. Its* Wodzicki residue *is defined by*

$$Res_W T =: \frac{1}{n(2\pi)^n} \int_{S^*M} tr_E \, \sigma_{-n}(T) d\mu \;. \quad (6.29)$$

Here, $\sigma_{-n}(T)$ is the principal symbol: a matrix-valued function on T^*M which is homogeneous of degree $-n$ in the fibre coordinates (see App. A.6). The integral is taken over the unit co-sphere

$$S^*M = \{(x,\xi) \in T^*M \mid \|\xi\| = 1\} \subset T^*M \;, \quad (6.30)$$

with measure $d\mu = dx d\xi$. The trace tr_E is a matrix trace over 'internal indices'.[1]

Example 6.3.1. Powers of the Laplacian on the n-dimensional flat torus T^n. The Laplacian Δ is a second order operator. Then, the operator $\Delta^{-n/2}$ is of order $-n$ with principal symbol $\sigma_{-n}(\Delta^{-n/2}) = \|\xi\|^{-n}$ (see App. A.6), which is the constant function 1 on S^*T^n. As a consequence,

$$
\begin{aligned}
Res_W \Delta^{-n/2} &= \frac{1}{n(2\pi)^n} \int_{S^*T^n} dx d\xi \\
&= \frac{1}{n(2\pi)^n} \Omega_n \int_{S^*T^n} dx \\
&= \frac{2\pi^{n/2}}{n\Gamma(n/2)} \;. \quad (6.31)
\end{aligned}
$$

The result coincides with the one given by the Dixmier trace in Example 6.2.1.

---

[1] It may be worth mentioning that most authors do not include the factor $\frac{1}{n}$ in the definition of the residue (6.29).

Example 6.3.2. Powers of the Laplacian on the n-dimensional sphere S^n. Again the operator $\Delta^{-n/2}$ is of order $-n$ with principal symbol the constant function 1 on S^*T^n. Thus,

$$
\begin{aligned}
Res_W \Delta^{-n/2} &= \frac{1}{n(2\pi)^n} \int_{S^*S^n} dx d\xi \\
&= \frac{1}{n(2\pi)^n} \Omega_n \int_{S^*T^n} dx \\
&= \frac{1}{n(2\pi)^n} \Omega_n \Omega_{n+1} \\
&= \frac{2\pi^{n/2}}{n\Gamma(n/2)} = \frac{2}{n!} ,
\end{aligned} \tag{6.32}
$$

where we have used the formula $\Gamma(\frac{n}{2})\Gamma(\frac{n+1}{2}) = 2^{-n+1}\pi^{1/2}(n-1)!$. Again we see that the result coincides with the one in Example 6.2.2 obtained by taking the Dixmier trace.

Example 6.3.3. The inverse of the one dimensional harmonic oscillator. The Hamiltonian is given by $H = \frac{1}{2}(\xi^2 + x^2)$. Let us forget for the moment the fact that the manifold we are considering, $M = \mathbb{R}$, is not compact. We would like to still make sense of the (Wodzicki) residue of a suitable negative power of H. Since H is of order 2, the first candidate would be $H^{-1/2}$. From (A.84) its principal symbol is the function ξ^{-1}. Formula (6.29) would give $Res_W H^{-1/2} = \infty$, a manifestation of the fact that $\mathbb{R}$ is not compact. On the other hand, Example 6.2.3 would suggest we try H^{-1}. But from (A.84) we see that the symbol of H^{-1} has no term of order -1 ! It is somewhat surprising that the integral of the *full* symbol of H^{-1} gives an answer which coincides (up to a factor 2) with $tr_\omega(H^{-1})$ evaluated in Example 6.2.3,

$$
\text{Residue}(H^{-1}) = \frac{1}{2\pi} \int_{S^*\mathbb{R}} \sigma(H^{-1}) = \frac{1}{\pi} \int_{\mathbb{R}} \frac{2}{1+x^2} = 2 . \tag{6.33}
$$

For an explanation of the previous fact we refer to [63].

As we have already mentioned, Wodzicki [134] has extended the formula (6.29) to a unique trace on the algebra of pseudodifferential operator of any order. The trace of any operator T is given by the right hand side of formula (6.29), with $\sigma_{-n}(T)$ the symbol of order $-n$ of T. In particular, one puts $Res_W T = 0$ if the order of T is less than $-n$. For additional material we refer to [85, 64].

In the examples worked out above we have explicitly seen that the Dixmier trace of an operator of a suitable type coincides with the Wodzicki residue of the operator. That the residue coincides with the Dixmier trace for any pseudodifferential operators of order less or equal that $-n$ has been shown by Connes [31, 32] (see also [130]).

Proposition 6.3.1. *Let M be an n dimensional compact Riemannian manifold. Let T be a pseudodifferential operator of order $-n$ acting on sections of a complex vector bundle $E \to M$.*
Then,

1. *The corresponding operator T on the Hilbert space $\mathcal{H} = L^2(M, E)$ of square integrable sections, belongs to $\mathcal{L}^{(1,\infty)}$.*
2. *The trace $tr_\omega T$ does not depend on ω and coincides with the residue,*

$$tr_\omega T = Res_W T =: \frac{1}{n(2\pi)^n} \int_{S^*M} tr_E \; \sigma_{-n}(T) d\mu \; . \qquad (6.34)$$

3. *The trace depends only on the conformal class of the metric on M.*

Proof. The Hilbert space on which T acts is just $\mathcal{H} = L^2(M, E)$, the space of square-integrable sections obtained as the completion of $\Gamma(M, E)$ with respect to the scalar product $(u_1, u_2) = \int_M u_1^* u_2 d\mu(g)$, $d\mu(g)$ being the measure associated with the Riemannian metric on M. If $\mathcal{H}_1, \mathcal{H}_2$ are obtained from two conformally related metrics, the identity operator on $\Gamma(M, E)$ extends to a linear map $U : \mathcal{H}_1 \to \mathcal{H}_2$ which is bounded with bounded inverse and which transforms T into UTU^{-1}. Since $tr_\omega(UTU^{-1}) = tr_\omega(T)$, we get $\mathcal{L}^{(1,\infty)}(\mathcal{H}_1) \simeq \mathcal{L}^{(1,\infty)}(\mathcal{H}_2)$ and the Dixmier trace does not change. On the other side, the cosphere bundle S^*M is constructed by using a metric. But since $\sigma_{-n}(T)$ is homogeneous of degree $-n$ in the fibre variable ξ, the multiplicative term obtained by changing variables just compensates the Jacobian of the transformation. Thus the integral in the definition of the Wodzicki residue remains the same in each conformal class.

Now, as we shall see in App. A.6, any operator T can be written as a finite sum of operators of the form $u \mapsto \phi T \psi$, with ϕ, ψ belonging to a partition of unity of M. Since multiplication operators are bounded on the Hilbert space $\mathcal{H}$, the operator T will be in $\mathcal{L}^{(1,\infty)}$ if and only if all operators $\phi T \psi$ are. Thus one can assume that E is the trivial bundle and M can be taken to be a given n-dimensional compact manifold, $M = S^n$ for simplicity. Now, it turns out that the operator T can be written as $T = S(1 + \Delta)^{-n/2}$, with Δ the Laplacian and S a bounded operator. From Example 6.2.2, we know that $(1 + \Delta)^{-n/2} \in \mathcal{L}^{(1,\infty)}$, (the presence of the identity is irrelevant since it produces only terms of lower degree), and this implies that $T \in \mathcal{L}^{(1,\infty)}$. From that example, we also have that for $s < -n/2$, the Dixmier trace of $(1 + \Delta)^s$ vanishes and this implies that any pseudodifferential operator on M of order $s < -n/2$ has vanishing Dixmier trace. In particular, the operator of order $(-n - 1)$ whose symbol is $\sigma(x, \xi) - \sigma_{-n}(x, \xi)$ has vanishing Dixmier trace; as a consequence, the Dixmier trace of T depends only on the principal symbol of T.

Now, the space of all $tr_E s_{-n}(T)$ can be identified with $C^\infty(S^*M)$. Furthermore, the map $tr_E s_{-n}(T) \mapsto tr_\omega(T)$ is a continuous linear form, namely a distribution, on the compact manifold S^*M. This distribution is positive due to the fact that the Dixmier trace is a positive linear functional and non-negative principal symbols correspond to positive operators. Since a positive distribution is a measure dm, we can write $tr_\omega(T) = \int_{S^*M} \sigma_{-n}(T)dm(x, \xi)$. Now, an isometry $\phi : S^n \to S^n$ will transform the symbol $\sigma_{-n}(T)(x, \xi)$ to $\sigma_{-n}(T)(\phi(x), \phi^*\xi)$, ϕ^* being the transpose of the Jacobian of ϕ, and determines a unitary operator U_ϕ on $\mathcal{H}$ which transforms T to $U_\phi T U_\phi^{-1}$. Since $tr_\omega T = tr_\omega(U_\phi T U_\phi^{-1})$, the measure dm determined by tr_ω is invariant under all isometries of S^n. In particular one can take $\phi \in SO(n+1)$. But S^*S^n is a homogeneous space for the action of $SO(n+1)$ and any $SO(n+1)$-invariant measure is proportional to the volume form on S^*S^n. Thus

$$tr_\omega T \sim \frac{1}{n(2\pi)^n} \int_{S^*M} tr_E \, \sigma_{-n}(T)dxd\xi \; = Res_W T \, . \tag{6.35}$$

From Examples 6.2.2 and 6.3.2 we see that the proportionally constant is just 1. This ends the proof of the proposition.

Finally, we mention that in general there is a class $\mathcal{M}$ of elements of $\mathcal{L}^{(1,\infty)}$ for which the Dixmier trace does not depend on the functional ω. Such operators are called *measurable* and in all the relevant cases in noncommutative geometry one deals with measurable operators. We refer to [32] for a characterization of $\mathcal{M}$. We only mention that in such situations, the Dixmier trace can again be written as a residue. If T is a positive element in $\mathcal{L}^{(1,\infty)}$, its complex power $T^s, s \in \mathbb{C}$, $\mathbb{R}e \; s > 1$, makes sense and is a trace class operator. Its trace

$$\zeta(s) = tr \, T^s = \sum_{n=0}^{\infty} \mu_n(T)^s \, , \tag{6.36}$$

is a holomorphic function on the half plane $\mathbb{R}e \; s > 1$. Connes has proven that for T a positive element in $\mathcal{L}^{(1,\infty)}$, the limit $\lim_{s\to 1+}(s-1)\zeta(s) = L$ exists if and only if

$$tr_\omega T =: \lim_{N\to\infty} \frac{1}{\ln N} \sum_0^{N-1} \mu_n(T) = L \, . \tag{6.37}$$

We see that if $\zeta(s)$ has a simple pole at $s = 1$ then, the corresponding residue coincides with the Dixmier trace. This equality gives back Proposition 6.3.1 for pseudodifferential operators of order at most $-n$ on a compact manifold of dimension n.

In Sect. 10.1 we shall describe the use of such general Wodzicki residues in the construction of gravity theories in the framework of noncommutative geometry. Here we only mention that the residue has found other applications in physics. In [104] it has been used to compute the Schwinger terms for current algebras, while in [57], in the context of zeta-function regularization, it has been used to evaluate all residues of zeta functions $\zeta(s)$ and to evaluate the multiplicative anomaly of regularized determinants of products of pseudo differential operators.

6.4 Spectral Triples

We shall now illustrate the basic ingredient introduced by Connes to develop the analogue of differential calculus for noncommutative algebras.

Definition 6.4.1. *A spectral triple $(\mathcal{A}, \mathcal{H}, D)$[2] is given by an involutive algebra $\mathcal{A}$ of bounded operators on the Hilbert space $\mathcal{H}$, together with a self-adjoint operator $D = D^*$ on $\mathcal{H}$ with the following properties.*

1. The resolvent $(D - \lambda)^{-1}$, $\lambda \notin \mathbb{R}$, is a compact operator on $\mathcal{H}$;
2. $[D, a] =: Da - aD \in \mathcal{B}(\mathcal{H})$, for any $a \in \mathcal{A}$.

The triple is said to be even *if there is a $\mathbb{Z}_2$ grading of $\mathcal{H}$, namely an operator Γ on $\mathcal{H}$, $\Gamma = \Gamma^*, \Gamma^2 = 1$, such that*

$$\Gamma D + D\Gamma = 0 \ ,$$
$$\Gamma a - a\Gamma = 0 \ , \quad \forall \, a \in \mathcal{A} \ . \tag{6.38}$$

If such a grading does not exist, the triple is said to be odd.

By the assumptions in Definition 6.4.1, the self-adjoint operator D has a real discrete spectrum made of eigenvalues, i.e. the collection $\{\lambda_n\}$ forms a discrete subset of $\mathbb{R}$, and each eigenvalue has finite multiplicity. Furthermore, $|\lambda_n| \to \infty$ as $n \to \infty$. Indeed, $(D - \lambda)^{-1}$ being compact, has characteristic values $\mu_n((D - \lambda)^{-1}) \to 0$, from which $|\lambda_n| = \mu_n(|D|) \to \infty$.

In general, one could ask that condition 2. of Definition 6.4.1 be satisfied only for a dense subalgebra of $\mathcal{A}$.

Various degrees of regularity of elements of $\mathcal{A}$ are defined using D and $|D|$. The reason for the corresponding names will be evident in the next subsection where we shall consider the canonical triple associated with an ordinary manifold. To start with, $a \in \mathcal{A}$ will be said to be *Lipschitz* if and only if the commutator $[D, a]$ is bounded. As mentioned before, in general this condition selects a dense subalgebra of $\mathcal{A}$. We recall that, if $\mathcal{D} \subset \mathcal{H}$ denotes the (dense) domain of D, the condition $[D, a] \in \mathcal{B}(\mathcal{H})$ can be expressed in one of the following equivalent ways [19]:

1. The element a is in the domain of the derivation $[D, \cdot \,]$ of $\mathcal{B}(\mathcal{H})$, which is the generator of the 1-parameter group σ_s of automorphism of $\mathcal{B}(\mathcal{H})$ given by

$$\sigma_s(T) = e^{isD} T e^{-isD} \ , \quad T \in \mathcal{B}(\mathcal{H}) \ . \tag{6.39}$$

2. For any $\varphi, \psi \in \mathcal{D}$, the sesquilinear form

$$q(\varphi, \psi) =: (D\varphi, a\psi)_{\mathcal{H}} - (a^*\varphi, D\psi)_{\mathcal{H}} \ , \tag{6.40}$$

is bounded on $\mathcal{D} \times \mathcal{D}$.

---

[2] The couple $(\mathcal{H}, D)$ is also called a K-cycle over $\mathcal{A}$.

3. For any $\varphi \in \mathcal{D}$ one has that $a\varphi \in \mathcal{D}$ and the commutator $[D, a]$ is norm bounded on $\mathcal{D}$.

Furthermore, consider the derivation δ on $\mathcal{B}(\mathcal{H})$ defined by

$$\delta(T) = [|D|, T], \quad T \in \mathcal{B}(\mathcal{H}). \tag{6.41}$$

It is the generator of the 1-parameter group α_s of automorphism of $\mathcal{B}(\mathcal{H})$ given by

$$\alpha_s(T) = e^{is|D|} T e^{-is|D|}. \tag{6.42}$$

Given the derivation δ, one defines the subalgebra $\mathcal{A}^k \subset \mathcal{A}$, with $k \geq 2$, as the one generated by elements $a \in \mathcal{A}$ such that both a and $[D, a]$ are in the domain of δ^{k-1}. One could also think of $\mathcal{A}^0$ as being just $\mathcal{A}$ and of $\mathcal{A}^1$ as consisting of the Lipschitz elements. The element $a \in \mathcal{A}$ is said to be of class C^∞ if it belongs to $\bigcap_{k \in \mathbb{N}} \mathcal{A}^k$.

It is worth mentioning that higher order commutators with D cannot be used to express higher order regularity conditions. As we shall see in the next Section, in the commutative situation, while $[D, f]$ is a multiplicative operator and therefore is bounded, $[D, [D, f]]$ is the sum of a multiplicative operator and of a differential operator (see footnote on page 97) and therefore it is not bounded. On the other hand, in the commutative framework δ^k is bounded on both C^∞ functions and forms.

In Sect. 7.3 the subalgebra $\mathcal{A}^2$ will play a crucial rôle in the definition of a scalar product on noncommutative forms.

We end this Section by mentioning that, as will be evident from the next Section, the spectral triples we are considering are really 'Euclidean' ones. There are some attempts to construct spectral triples with 'Minkowskian signature' [81, 73, 90]. We shall not use them in these notes.

6.5 The Canonical Triple over a Manifold

The basic example of spectral triple is constructed by means of the Dirac operator on a closed n-dimensional Riemannian spin[3] manifold (M, g). The corresponding spectral triple $(\mathcal{A}, \mathcal{H}, D)$ will be called the *canonical triple* over the manifold M. For its constituents one takes:

1. $\mathcal{A} = \mathcal{F}(M)$ is the algebra of complex valued smooth functions on M.

---

[3] For much of what follows one could consider spin^c manifolds. The obstruction for a manifold to have a spin^c structure is rather mild and much weaker than the obstruction to having a spin structure. For instance, any orientable four dimensional manifold admits such a spin^c structure [5]. Then, one should accordingly modify the Dirac operator in (6.52) by adding a $U(1)$ gauge connection $A = dx^\mu A_\mu$. The corresponding Hilbert space $\mathcal{H}$ has a beautiful interpretation as the space of square integrable Pauli-Dirac spinors [67].

2. $\mathcal{H} = L^2(M, S)$ is the Hilbert space of square integrable sections of the irreducible spinor bundle over M; its rank being equal to $2^{[n/2]}$.[4] The scalar product in $L^2(M, S)$ is the usual one of the measure $d\mu(g)$ associated with the metric g,

$$(\psi, \phi) = \int d\mu(g)\overline{\psi(x)}\phi(x), \tag{6.43}$$

with bar indicating complex conjugation and scalar product in the spinor space being the natural one in $\mathbb{C}^{2^{[n/2]}}$.

3. D is the Dirac operator associated with the Levi-Civita connection $\omega = dx^\mu \omega_\mu$ of the metric g.[5]

First of all, notice that the elements of the algebra $\mathcal{A}$ act as multiplicative operators on $\mathcal{H}$,

$$(f\psi)(x) =: f(x)\psi(x) , \quad \forall f \in \mathcal{A} , \psi \in \mathcal{H} . \tag{6.44}$$

Next, let $(e_a, a = 1, \ldots, n)$ be an orthonormal basis of vector fields which is related to the natural basis $(\partial_\mu, \mu = 1, \ldots, n)$ via the n-beins, with components e_a^μ, so that the components $\{g^{\mu\nu}\}$ and $\{\eta^{ab}\}$ of the curved and the flat metrics respectively, are related by,

$$g^{\mu\nu} = e_a^\mu e_b^\nu \eta^{ab} , \quad \eta_{ab} = e_a^\mu e_b^\nu g_{\mu\nu} . \tag{6.45}$$

From now on, the curved indices $\{\mu\}$ and the flat ones $\{a\}$ will run from 1 to n and as usual we sum over repeated indices. Curved indices will be lowered and raised by the curved metric g, while flat indices will be lowered and raised by the flat metric η.

The coefficients $(\omega_{\mu a}{}^b)$ of the Levi-Civita (metric and torsion-free) connection of the metric g, defined by $\nabla_\mu e_a = \omega_{\mu a}{}^b e_b$, are the solutions of the equations

$$\partial_\mu e_\nu^a - \partial_\nu e_\mu^a - \omega_{\mu b}{}^a e_\nu^b + \omega_{\nu b}{}^a e_\mu^b = 0 . \tag{6.46}$$

Also, let $C(M)$ be the Clifford bundle over M whose fiber at $x \in M$ is just the complexified Clifford algebra $Cliff_\mathbb{C}(T_x^*M)$ and $\Gamma(M, C(M))$ be the module of corresponding sections. We have an algebra morphism

$$\gamma : \Gamma(M, C(M)) \to \mathcal{B}(\mathcal{H}) , \tag{6.47}$$

defined by

$$\gamma(dx^\mu) =: \gamma^\mu(x) = \gamma^a e_a^\mu , \quad \mu = 1, \ldots, n , \tag{6.48}$$

and extended as an algebra map and by requiring $\mathcal{A}$-linearity.
The curved and flat gamma matrices $\{\gamma^\mu(x)\}$ and $\{\gamma^a\}$, which we take to be Hermitian, obey the relations

---

[4] The symbol $[k]$ indicates the integer part in k.
[5] To simplify matters, the kernel of the Dirac operator D is assumed to be trivial.

$$\gamma^\mu(x)\gamma^\nu(x) + \gamma^\nu(x)\gamma^\mu(x) = -2g(dx^\mu, dx^n) = -2g^{\mu\nu} \; , \quad \mu, \nu = 1, \ldots, n \; ;$$
$$\gamma^a\gamma^b + \gamma^b\gamma^a = -2\eta^{ab} \; , \quad a, b = 1, \ldots, n \; . \tag{6.49}$$

The lift ∇^S of the Levi-Civita connection to the bundle of spinors is then

$$\nabla^S_\mu = \partial_\mu + \omega^S_\mu = \partial_\mu + \frac{1}{4}\omega_{\mu ab}\gamma^a\gamma^b \; . \tag{6.50}$$

The Dirac operator, defined by

$$D = \gamma \circ \nabla^S \; , \tag{6.51}$$

can be written locally as

$$D = \gamma(dx^\mu)\nabla^S_\mu = \gamma^\mu(x)(\partial_\mu + \omega^S_\mu) = \gamma^a e^\mu_a(\partial_\mu + \omega^S_\mu) \; . \tag{6.52}$$

Finally, we mention the Lichnérowicz formula for the square of the Dirac operator [10],

$$D^2 = \Delta^S + \frac{1}{4}R \; . \tag{6.53}$$

Here R is the scalar curvature of the metric and Δ^S is the Laplacian operator lifted to the bundle of spinors,

$$\Delta^S = -g^{\mu\nu}(\nabla^S_\mu\nabla^S_\nu - \Gamma^\rho_{\mu\nu}\nabla^S_\rho) \; , \tag{6.54}$$

with $\Gamma^\rho_{\mu\nu}$ the Christoffel symbols of the connection.

If the dimension n of M is even, the previous spectral triple is even. For the grading operator one just takes the product of all flat gamma matrices,

$$\Gamma = \gamma^{n+1} = i^{n/2}\gamma^1 \cdots \gamma^n \; , \tag{6.55}$$

which, since n is even, anticommutes with the Dirac operator,

$$\Gamma D + D\Gamma = 0 \; . \tag{6.56}$$

Furthermore, the factor $i^{n/2}$ ensures that

$$\Gamma^2 = \mathbb{I} \; , \quad \Gamma^* = \Gamma \; . \tag{6.57}$$

Proposition 6.5.1. *Let $(\mathcal{A}, \mathcal{H}, D)$ be the canonical triple over the manifold M as defined above. Then*

1. *The space M is the structure space of the algebra $\overline{\mathcal{A}}$ of continuous functions on M, which is the norm closure of $\mathcal{A}$.*
2. *The geodesic distance between any two points on M is given by*

$$d(p, q) = \sup_{f \in \mathcal{A}}\{|f(p) - f(q)| \mid \|[D, f]\| \leq 1\} \; , \quad \forall \, p, q \in M \; . \tag{6.58}$$

3. *The Riemannian measure on M is given by*

$$\int_M f \;=\; c(n)\; tr_\omega(f|D|^{-n}) \;,\quad \forall\, f \in \mathcal{A}\;,$$

$$c(n) = 2^{(n-[n/2]-1)}\pi^{n/2} n\Gamma(\frac{n}{2})\;. \tag{6.59}$$

Proof. Statement 1. is just the Gel'fand-Naimark theorem illustrated in Sect. 2.2.

As for Statement 2., from the action (6.44) of $\mathcal{A}$ as multiplicative operators on $\mathcal{H}$, one finds that

$$[D, f]\psi = (\gamma^\mu \partial_\mu f)\psi\;,\quad \forall\, f \in \mathcal{A}\;, \tag{6.60}$$

and the commutator $[D, f]$ is a multiplicative operator as well[6] ,

$$[D, f] = \gamma^\mu \partial_\mu f = \gamma(df)\;,\quad \forall\, f \in \mathcal{A}\;. \tag{6.61}$$

As a consequence, its norm is

$$\||[D, f]\|| = sup|(\gamma^\mu \partial_\mu f)(\gamma^\nu \partial_\nu f)^*|^{1/2} = sup|\gamma^{\mu\nu}\partial_\mu f \partial_\nu f^*|^{1/2}\;. \tag{6.62}$$

Now, the right-hand side of (6.62) coincides with the Lipschitz norm of f [32] which is given by

$$\|f\|_{Lip} =: \sup_{x \neq y} \frac{|f(x) - f(y)|}{d_\gamma(x, y)}\;, \tag{6.63}$$

with d_γ the usual geodesic distance on M, given by the usual formula,

$$d_\gamma(x, y) = inf_\gamma\{\text{ length of paths } \gamma \text{ from } x \text{ to } y\ \}\;. \tag{6.64}$$

Therefore, we have that

$$\||[D, f]\|| = \sup_{x \neq y} \frac{|f(x) - f(y)|}{d_\gamma(x, y)}\;. \tag{6.65}$$

Now, the condition $\||[D, f]\|| \leq 1$ in (6.58), automatically gives

$$d(p, q) \leq d_\gamma(p, q)\;. \tag{6.66}$$

To invert the inequality sign, fix the point q and consider the function $f_{\gamma,q}(x) = d_\gamma(x, q)$. Then $\||[D, f_{\gamma,q}]\|| \leq 1$, and in (6.58) this gives

$$d(p, q) \geq |f_{\gamma,q}(p) - f_{\gamma,q}(q)| = d_\gamma(p, q)\;, \tag{6.67}$$

which, together with (6.66) proves Statement 2. As a very simple example, consider $M = \mathbb{R}$ and $D = \frac{d}{dx}$. Then, the condition $\||[D, f]\|| \leq 1$ is just

---

[6] One readily finds that $[D, [D, f]] = \gamma^\mu\gamma^\nu\partial_\mu\partial_\nu f + (\gamma^\mu\gamma^\nu - \gamma^\nu\gamma^\mu)(\partial_\nu f)\partial_\mu$ which, being a sum of a multiplicative operator and a differential operator cannot be bounded. This is the reason why higher commutators are not used for defining higher order regularity conditions for functions (see Sect. 6.4).

$\sup |\frac{df}{dx}| \le 1$ and the sup is saturated by the function $f(x) = x + cost$ which gives the usual distance.

The proof of Statement 3. starts with the observation that the principal symbol of the Dirac operator is $\gamma(\xi)$, left multiplication by $\gamma(\xi)$ on spinor fields, and so D is a first-order elliptic operator (see App. A.6). Since any $f \in \mathcal{A}$ acts as a bounded multiplicative operator, the operator $f|D|^{-n}$ is pseudodifferential of order $-n$. Its principal symbol is $\sigma_{-n}(x, \xi) = f(x)\|\xi\|^{-n}$ which, on the co-sphere bundle $\|\xi\| = 1$, reduces to the matrix $f(x)\mathbb{I}_{2^{[n/2]}}$, $2^{[n/2]} = dimS_x$, S_x being the fibre of S. From the trace theorem, Prop 6.3.1, we get

$$
\begin{aligned}
tr_\omega(f|D|^{-n}) &= \frac{1}{n(2\pi)^n} \int_{S^*M} tr(f(x)\mathbb{I}_{2^{[n/2]}})dxd\xi \\
&= \frac{2^{[n/2]}}{n(2\pi)^n}\left(\int_{S^{n-1}} d\xi\right) \int_M f(x)dx \\
&= \frac{1}{c(n)} \int_M f \, .
\end{aligned}
\tag{6.68}
$$

Here, $\int_{S^{n-1}} d\xi = 2\pi^{n/2}/\Gamma(n/2)$ is the area of the unit sphere S^{n-1}. This gives $c(n) = 2^{(n-[n/2]-1)}\pi^{n/2}n\Gamma(n/2)$ and Statement 3. is proven.

It is worth mentioning that the geodesic distance (6.58) can also be recovered from the Laplace operator ∇_g associated with the Riemannian metric g on M [66, 67]. One has that

$$
d(p, q) = \sup_f\{|f(p) - f(q)| \mid \|f\nabla f - \frac{1}{2}(\nabla f^2 + f^2\nabla)\|_{L^2(M)} \le 1\} , \quad (6.69)
$$

where $L^2(M)$ is the Hilbert space of square integrable *functions* on M. Indeed, the operator $f\nabla f - \frac{1}{2}(\nabla f^2 + f^2\nabla)$ is just the multiplicative operator $g^{\mu\nu}\partial_\mu f\partial_\nu f$. Thus, much of the usual differential geometry can be recovered from the triple $(C^\infty(M), L^2(M), \nabla_g)$, although it is technically much more involved.

6.6 Distance and Integral for a Spectral Triple

Given a general spectral triple $(\mathcal{A}, \mathcal{H}, D)$, there is an analogue of formula (6.58) which gives a natural distance function on the space $S(\overline{\mathcal{A}})$ of states on the C^*-algebra $\overline{\mathcal{A}}$ (the norm closure of $\mathcal{A}$). A state on $\overline{\mathcal{A}}$ is any linear map $\phi : \mathcal{A} \to \mathbb{C}$ which is positive, i.e. $\phi(a^*a) > 0$, and normalized, i.e. $\phi(\mathbb{I}) = 1$ (see also App. A.2). The distance function on $S(\overline{\mathcal{A}})$ is defined by

$$
d(\gamma, \chi) =: \sup_{a \in \mathcal{A}}\{|\phi(a) - \chi(a)| \mid \|[D, a]\| \le 1\} , \quad \forall \, \phi, \chi \in S(\overline{\mathcal{A}}) . \tag{6.70}
$$

In order to define the analogue of the measure integral, one needs the additional notion of the dimension of a spectral triple.

Definition 6.6.1. *A spectral triple* $(A, \mathcal{H}, D)$ *is said to be of dimension* $n > 0$ *(or n summable) if* $|D|^{-1}$ *is an infinitesimal (in the sense of Definition 6.1.1) of order* $\frac{1}{n}$ *or, equivalently, if* $|D|^{-n}$ *is an infinitesimal or order* 1.

Having such a n-dimensional spectral triple, the *integral* of any $a \in A$ is defined by

$$\int a =: \frac{1}{V} tr_\omega a |D|^{-n} , \tag{6.71}$$

where the constant V is determined by the behavior of the characteristic values of $|D|^{-n}$, namely, $\mu_j \leq V j^{-1}$ for $j \to \infty$. We see that the rôle of the operator $|D|^{-n}$ is just to bring the bounded operator a into $\mathcal{L}^{(1,\infty)}$ so that the Dixmier trace makes sense. By construction, the integral in (6.71) is normalized,

$$\int \mathbb{I} = \frac{1}{V} tr_\omega |D|^{-n} = \frac{1}{V} \lim_{N \to \infty} \sum_{j=1}^{N-1} \mu_j(|D|^{-n}) = \lim_{N \to \infty} \sum_{j=1}^{N-1} \frac{1}{j} = 1 . \tag{6.72}$$

The operator $|D|^{-n}$ is the analogue of the volume of the space.
In Sect. 7.3 it will be shown that the integral (6.71) is a non-negative (normalized) trace on A, satisfying the following relations,

$$\int ab = \int ba , \quad \forall \, a, b \in A ,$$
$$\int a^* a \geq 0 , \quad \forall \, a \in A . \tag{6.73}$$

For the canonical spectral triple over a manifold M, its dimension coincides with the dimension of M. Indeed, the Weyl formula for the eigenvalues gives for large j [72],

$$\mu_j(|D|) \sim 2\pi (\frac{n}{\Omega_n vol M})^{1/n} j^{1/n} , \tag{6.74}$$

n being the dimension of M.

6.7 Real Spectral Triples

In fact, one needs to introduce an additional notion: the *real structure* one. The latter is essential to introduce Poincaré duality and plays a crucial rôle in the derivation of the Lagrangian of the Standard Model [34, 36]. This real structure may be thought of as a generalized CPT operator (in fact only CP, since we are taking Euclidean signature).

Definition 6.7.1. *Let* $(A, \mathcal{H}, D)$ *be a spectral triple of dimension n. A real structure is an antilinear isometry* $J : \mathcal{H} \to \mathcal{H}$, *with the properties*

1a. $\quad J^2 = \varepsilon(n)\mathbb{I}$,
1b. $\quad JD = \varepsilon'(n)DJ$,
1c. $\quad J\Gamma = (i)^n\Gamma J$; *if n is even with Γ the $\mathbb{Z}_2$-grading.*
2a. $\quad [a, b^0] = 0$,
2b. $\quad [[D, a], b^0] = 0$, $\quad b^0 = Jb^*J^*$, $\quad$ *for any $a, b \in \mathcal{A}$* .

The mod 8 periodic functions $\varepsilon(n)$ and $\varepsilon'(n)$ are given by [34]

$$\varepsilon(n) = (1, 1, -1, -1, -1, -1, 1, 1) ,$$
$$\varepsilon'(n) = (1, -1, 1, 1, 1, -1, 1, 1) , \tag{6.75}$$

n being the dimension of the triple. The previous periodicity is a manifestation of the so called 'spinorial chessboard' [23].

A full analysis of the previous conditions goes beyond the scope of these notes. We only mention that 2a. is used by Connes to formulate Poincaré duality and to define noncommutative manifolds. The map J is related to the Tomita(-Takesaki) involution. The Tomita theorem states that for any weakly closed[7] *-algebra of operators $\mathcal{M}$ on a Hilbert space $\mathcal{H}$ which admits a cyclic and separating vector[8], there exists a canonical antilinear isometric involution $J : \mathcal{H} \to \mathcal{H}$ which conjugates $\mathcal{M}$ to its commutant

$$\mathcal{M}' =: \{T \in \mathcal{B}(\mathcal{H}) \mid Ta = aT , \quad \forall \, a \in \mathcal{M}\} , \tag{6.76}$$

i.e. $JMJ^* = \mathcal{M}'$. As a consequence, $\mathcal{M}$ is anti-isomorphic to $\mathcal{M}'$, the anti-isomorphism being given by the map $\mathcal{M} \ni a \mapsto Ja^*J^* \in \mathcal{M}'$. The existence of the map J satisfying condition 2a. also turns the Hilbert space $\mathcal{H}$ into a bimodule over $\mathcal{A}$, the bimodule structure being given by

$$a \, \xi \, b =: aJb^*J^* \, \xi , \quad \forall \, a, b \in \mathcal{A} . \tag{6.77}$$

As for condition 2b., for the time being, it may be thought of as the statement that D is a 'generalized differential operator' of order 1. As we shall see, it will play a crucial rôle in the spectral geometry described in Sect. 9.3. It is worth stressing that, since a and b^0 commute by condition 2a., condition 2b. is symmetric, namely it is equivalent to the condition $[[D, b^0], a] = 0$, for any $a, b \in \mathcal{A}$.
If $a \in \mathcal{A}$ acts on $\mathcal{H}$ as a *left* multiplication operator, then Ja^*J^* is the corresponding *right* multiplication operator. For commutative algebras, these actions can be identified and one simply writes $a = Ja^*J^*$. Then, condition 2b. reads $[[D, a], b] = 0$, for any $a, b \in \mathcal{A}$, which is just the statement that D is a differential operator of order 1.

---

[7] We recall that the sequence $\{T_\lambda\}_{\lambda \in \Lambda}$ is said to converge *weakly to T, $T_\lambda \to T$,* if and only if, for any $\xi, \eta \in \mathcal{H}$, $\langle (T_\lambda - T)\xi, \eta \rangle \to 0$.

[8] If $\mathcal{M}$ is an involutive subalgebra of $\mathcal{B}(\mathcal{H})$, a vector $\xi \in \mathcal{H}$ is called *cyclic* for $\mathcal{M}$ if $\mathcal{M}\xi$ is dense in $\mathcal{H}$. It is called *separating* for $\mathcal{M}$ if for any $T \in \mathcal{M}$, $T\xi = 0$ implies $T = 0$. One finds that a cyclic vector for $\mathcal{M}$ is separating for the commutant $\mathcal{M}'$. If $\mathcal{M}$ is a von Neumann algebra ($\mathcal{M} = \mathcal{M}''$), the converse is also true, namely a cyclic vector for $\mathcal{M}'$ is separating for $\mathcal{M}$ [43].

The canonical triple associated with any (Riemannian spin) manifold has a canonical real structure in the sense of Definition 6.7.1, the antilinear isometry J being given by

$$J\psi =: C\overline{\psi} \ , \quad \forall \, \psi \in \mathcal{H} \ , \tag{6.78}$$

where C is the charge conjugation operator and bar indicates complex conjugation [23]. One verifies that all defining properties of J hold true.

6.8 A Two-Point Space

Consider a space made of two points $Y = \{1, 2\}$. The algebra $\mathcal{A}$ of continuous functions is the direct sum $\mathcal{A} = \mathbb{C} \oplus \mathbb{C}$ and any element $f \in \mathcal{A}$ is a couple of complex numbers (f_1, f_2), with $f_i = f(i)$ the value of f at the point i. A 0-dimensional even spectral triple $(\mathcal{A}, \mathcal{H}, D, \Gamma)$ is constructed as follows. The finite dimensional Hilbert space $\mathcal{H}$ is a direct sum $\mathcal{H} = \mathcal{H}_1 \oplus \mathcal{H}_2$ and elements of $\mathcal{A}$ act as diagonal matrices

$$\mathcal{A} \ni f \mapsto \left[\begin{array}{cc} f_1 \mathbb{I}_{dim H_1} & 0 \\ 0 & f_2 \mathbb{I}_{dim H_2} \end{array} \right] \in \mathcal{B}(\mathcal{H}) \ . \tag{6.79}$$

We shall identify every element of $\mathcal{A}$ with its matrix representation. The operator D can be taken as a 2×2 off-diagonal matrix, since any diagonal element would drop out of commutators with elements of $\mathcal{A}$,

$$D = \left[\begin{array}{cc} 0 & M^* \\ M & 0 \end{array} \right] \ , \quad M \in Lin(\mathcal{H}_1, \mathcal{H}_2) \ . \tag{6.80}$$

Finally, the grading operator Γ is given by

$$\Gamma = \left[\begin{array}{cc} \mathbb{I}_{dim H_1} & 0 \\ 0 & -\mathbb{I}_{dim H_2} \end{array} \right] \ . \tag{6.81}$$

With $f \in \mathcal{A}$, one finds for the commutator

$$[D, f] = (f_2 - f_1) \left[\begin{array}{cc} 0 & M^* \\ -M & 0 \end{array} \right] \ , \tag{6.82}$$

and, in turn, for its norm, $\|[D, f]\| = |f_2 - f_1| \lambda$ with λ the largest eigenvalue of the matrix $|M| = \sqrt{MM^*}$. Therefore, the noncommutative distance between the two points of the space is found to be

$$d(1, 2) = sup\{|f_2 - f_1| \mid \|[D, f]\| \leq 1\} = \frac{1}{\lambda} \ . \tag{6.83}$$

For the previous triple the Dixmier trace is just (a multiple of the) usual matrix trace.

A real structure J is given by

$$J \left(\begin{array}{c} \xi \\ \eta \end{array} \right) = \left(\begin{array}{c} \eta \\ \xi \end{array} \right) , \quad \forall \ (\xi, \eta) \in \mathcal{H}_1 \oplus \mathcal{H}_2 . \tag{6.84}$$

One checks that $J^2 = \mathbb{I}$, $\Gamma J + J\Gamma = 0$, $DJ - JD = 0$ and that all other requirements in the Definition 6.7.1 are satisfied.

6.9 Products and Equivalence of Spectral Triples

We shall briefly mention two additional concepts which are useful in general and in particular in the description of the Standard Model. These are the notions of product and equivalence of triples.

Suppose we have two spectral triples $(\mathcal{A}_1, \mathcal{H}_1, D_1, \Gamma_1)$ and $(\mathcal{A}_2, \mathcal{H}_2, D_2)$ the first one taken to be even with $\mathbb{Z}_2$-grading Γ_1 on $\mathcal{H}_1$. The product triple is the triple $(\mathcal{A}, \mathcal{H}, D)$ given by

$$\begin{aligned}
\mathcal{A} &=: \mathcal{A}_1 \otimes_{\mathbb{C}} \mathcal{A}_2 , \\
\mathcal{H} &=: \mathcal{H}_1 \otimes_{\mathbb{C}} \mathcal{H}_2 , \\
D &=: D_1 \otimes_{\mathbb{C}} \mathbb{I} + \Gamma_1 \otimes_{\mathbb{C}} D_2 .
\end{aligned} \tag{6.85}$$

From the definition of D and the fact that D_1 anticommutes with Γ_1 it follows that

$$\begin{aligned}
D^2 &= \frac{1}{2}\{D, D\} \\
&= (D_1)^2 \otimes_{\mathbb{C}} \mathbb{I} + (\Gamma_1)^2 \otimes_{\mathbb{C}} (D_2)^2 + \frac{1}{2}\{D_1, \Gamma_1\} \otimes_{\mathbb{C}} D_2 \\
&= (D_1)^2 \otimes_{\mathbb{C}} \mathbb{I} + \mathbb{I} \otimes_{\mathbb{C}} (D_2)^2 .
\end{aligned} \tag{6.86}$$

Thus, the dimensions sum up, namely, if D_j is of dimension n_j, that is $|D_j|^{-1}$ is an infinitesimal of order $1/n_j$, $j = 1, 2$, then D is of dimension $n_1 + n_2$, that is $|D|^{-1}$ is an infinitesimal of order $1/(n_1 + n_2)$. Furthermore, once the limiting procedure Lim_ω is fixed, one has also that [32],

$$\frac{\Gamma(n/2 + 1)}{\Gamma(n_1/2 + 1)\Gamma(n_2/2 + 1)} tr_\omega (T_1 \otimes T_2 |D|^n) = tr_\omega (T_1 |D|^{n_1}) tr_\omega (T_2 |D|^{n_2}) , \tag{6.87}$$

for any $T_j \in \mathcal{B}(\mathcal{H}_j)$. For the particular case in which one of the triples, the second one, say, is zero dimensional so that the Dixmier trace is the ordinary trace, the corresponding formula reads

$$tr_\omega (T_1 \otimes T_2 |D|^n) = tr_\omega (T_1 |D|^{n_1}) tr(T_2) . \tag{6.88}$$

The notion of equivalence of triples is the expected one. Suppose we are given two spectral triples $(\mathcal{A}_1, \mathcal{H}_1, D_1)$ and $(\mathcal{A}_2, \mathcal{H}_2, D_2)$, with the associated representations $\pi_j : \mathcal{A}_j \to \mathcal{B}(\mathcal{H}_j)$, $j = 1, 2$. Then, the triples are said to be *equivalent* if there exists a unitary operator $U : \mathcal{H}_1 \to \mathcal{H}_2$ such that

$U\pi_1(a)U^* = \pi_2(a)$ for any $a \in \mathcal{A}_1$, and $UD_1U^* = D_2$. If the two triples are even with grading operators Γ_1 and Γ_2 respectively, one requires also that $U\Gamma_1 U^* = \Gamma_2$. And if the two triples are real with real structure J_1 and J_2 respectively, one requires also that $UJ_1U^* = J_2$.

7. Noncommutative Differential Forms

We shall now describe how to construct a differential algebra of forms out of a spectral triple $(\mathcal{A}, \mathcal{H}, D)$. It turns out to be useful to first introduce a universal graded differential algebra which is associated with any algebra $\mathcal{A}$.

7.1 Universal Differential Forms

Let $\mathcal{A}$ be an associative algebra with unit (for simplicity) over the field of numbers $\mathbb{C}$ (say). The *universal differential algebra of forms* $\Omega\mathcal{A} = \bigoplus_p \Omega^p\mathcal{A}$ is a graded algebra defined as follows. In degree 0 it is equal to $\mathcal{A}$, $\Omega^0\mathcal{A} = \mathcal{A}$. The space $\Omega^1\mathcal{A}$ of *one-forms* is generated, as a left $\mathcal{A}$-module, by symbols of degree δa, for $a \in \mathcal{A}$, with relations

$$\delta(ab) = (\delta a)b + a\delta b , \quad \forall\, a, b \in \mathcal{A} . \tag{7.1}$$

$$\delta(\alpha a + \beta b) = \alpha\delta a + \beta\delta b , \quad \forall\, a, b \in \mathcal{A} , \quad \alpha, \beta \in \mathbb{C} . \tag{7.2}$$

We shall assume that the element $1 \in \mathbb{C}$ coincides with the unit of $\mathcal{A}$ (so that $\mathbb{C} \subset \mathcal{A}$) and that 1 is also the unit of the whole of $\Omega\mathcal{A}$. Then, the relation (7.1) automatically gives $\delta 1 = 0$, since $\delta 1 = \delta(1 \cdot 1) = (\delta 1) \cdot 1 + 1 \cdot (\delta 1) = 2(\delta 1)$ from which it follows that $\delta 1 = 0$. In turn this implies that $\delta\mathbb{C} = 0$.
A generic element $\omega \in \Omega^1\mathcal{A}$ is a finite sum of the form

$$\omega = \sum_i a_i \delta b_i , \quad a_i, b_i \in \mathcal{A} . \tag{7.3}$$

The left $\mathcal{A}$-module $\Omega^1\mathcal{A}$ can also be endowed with a structure of a right $\mathcal{A}$-module by

$$\left(\sum_i a_i \delta b_i\right)c =: \sum_i a_i(\delta b_i)c = \sum_i a_i \delta(b_i c) - \sum_i a_i b_i \delta c , \tag{7.4}$$

where, in the second equality, we have used (7.1). The relation (7.1) is just the Leibniz rule for the map

$$\delta : \mathcal{A} \to \Omega^1\mathcal{A} , \tag{7.5}$$

which can therefore be considered as a derivation of $\mathcal{A}$ with values in the bimodule $\Omega^1\mathcal{A}$. The pair $(\delta, \Omega^1\mathcal{A})$ is characterized by the following universal property [20, 24],

Proposition 7.1.1. *Let $\mathcal{M}$ be any $\mathcal{A}$-bimodule and $\Delta : \mathcal{A} \to \mathcal{M}$ any deriva-tion, namely any map which satisfies the rule (7.1). Then, there exists a unique bimodule morphism $\rho_\Delta : \Omega^1 \mathcal{A} \to \mathcal{M}$ such that $\Delta = \rho_\Delta \circ \delta$,*

$$id: \quad \Omega^1 \mathcal{A} \quad \longleftrightarrow \quad \Omega^1 \mathcal{A}$$

$$\delta \uparrow \qquad\qquad \downarrow \rho_\Delta \quad , \qquad \rho_\Delta \circ \delta = \Delta \ . \tag{7.6}$$

$$\Delta : \quad \mathcal{A} \quad \longrightarrow \quad \mathcal{M}$$

Proof. Notice, first of all, that for any bimodule morphism $\rho : \Omega^1 \mathcal{A} \to \mathcal{M}$ the composition $\rho \circ \delta$ is a derivation with values in $\mathcal{M}$. Conversely, let $\Delta : \mathcal{A} \to \mathcal{M}$ be a derivation; then, if there exists a bimodule morphism $\rho_\Delta : \Omega^1 \mathcal{A} \to \mathcal{M}$ such that $\Delta = \rho_\Delta \circ \delta$, it is unique. Indeed, the definition of δ gives

$$\rho_\Delta(\delta a) = \Delta(a) \ , \quad \forall \, a \in \mathcal{A} \ , \tag{7.7}$$

and the uniqueness follows from the fact that the image of δ generates $\Omega^1 \mathcal{A}$ as a left $\mathcal{A}$-module, if one extends the previous map by

$$\rho_\Delta(\sum_i a_i \delta b_i) = \sum_i a_i \Delta b_i \ , \quad \forall \, a_i, b_i \in \mathcal{A} \ . \tag{7.8}$$

It remains to prove that ρ_Δ as defined in (7.8) is a bimodule morphism. Now, with $a_i, b_i, f, g \in \mathcal{A}$, by using the fact that both δ and Δ are derivations, one has that

$$
\begin{aligned}
\rho_\Delta(f(\sum_i a_i \delta b_i)g) &= \rho_\Delta(\sum_i f a_i (\delta b_i)g) \\
&= \rho_\Delta(\sum_i f a_i \delta(b_i g) - \sum_i f a_i b_i \delta g) \\
&= \sum_i f a_i \Delta(b_i g) - \sum_i f a_i b_i \Delta g \\
&= \sum_i f a_i (\Delta b_i)g \\
&= f(\sum_i f a_i \Delta b_i)g \\
&= f(\sum_i a_i \Delta b_i)g \ ; \tag{7.9}
\end{aligned}
$$

and this ends the proof of the proposition.

Let us go back to universal forms. The space $\Omega^p \mathcal{A}$ is defined as

$$\Omega^p \mathcal{A} = \underbrace{\Omega^1 \mathcal{A} \Omega^1 \mathcal{A} \cdots \Omega^1 \mathcal{A} \Omega^1 \mathcal{A}}_{p-times} \ , \tag{7.10}$$

with the product of any two one-forms defined by 'juxtaposition',

$$(a_0 \delta a_1)(b_0 \delta b_1) \quad =: \quad a_0 (\delta a_1) b_0 \delta b_1$$
$$= \quad a_0 \delta (a_1 b_0) \delta b_1 - a_0 a_1 \delta b_0 \delta b_1 \, ,$$
$$\forall \, a_0, a_1, b_0, b_1 \in \mathcal{A} \, . \qquad (7.11)$$

Again we have used the rule (7.1). Thus, elements of $\Omega^p \mathcal{A}$ are finite linear combinations of monomials of the form

$$\omega = a_0 \delta a_1 \delta a_2 \cdots \delta a_p \, , \quad a_k \in \mathcal{A} \, . \qquad (7.12)$$

The product : $\Omega^p \mathcal{A} \times \Omega^q \mathcal{A} \to \Omega^{p+q} \mathcal{A}$ of any p-form with any q-form produces a $p+q$ form and is again defined by 'juxtaposition' and rearranging the result by using the relation (7.1),

$$(a_0 \delta a_1 \cdots \delta a_p)(a_{p+1} \delta a_{p+2} \cdots \delta a_{p+q}) =:$$
$$= a_0 \delta a_1 \cdots (\delta a_p) a_{p+1} \delta a_{p+2} \cdots \delta a_{p+q}$$
$$= (-1)^p a_0 a_1 \delta a_2 \cdots \delta a_{p+q}$$
$$+ \sum_{i=1}^{p} (-1)^{p-i} a_0 \delta a_1 \cdots \delta a_{i-1} \delta(a_i a_{i+1}) \delta a_{i+2} \cdots \delta a_{p+q} \, . \, (7.13)$$

The algebra $\Omega^p \mathcal{A}$ is a left $\mathcal{A}$-module by construction. It is also a right $\mathcal{A}$-module, the right structure being given by

$$(a_0 \delta a_1 \cdots \delta a_p) b \quad =: \quad a_0 \delta a_1 \cdots (\delta a_p) b$$
$$= \quad (-1)^p a_0 a_1 \delta a_2 \cdots \delta a_p \delta b$$
$$+ \sum_{i=1}^{p-1} (-1)^{p-i} a_0 \delta a_1 \cdots \delta a_{i-1} \delta(a_i a_{i+1}) \delta a_{i+2} \cdots \delta a_p \delta b$$
$$+ a_0 \delta a_1 \cdots \delta a_{p-1} \delta(a_p b) \, , \quad \forall \, a_i, b \in \mathcal{A} \, . \qquad (7.14)$$

Next, one makes the algebra $\Omega \mathcal{A}$ a differential algebra by 'extending' the *differential* δ to an operator : $\Omega^p \mathcal{A} \to \Omega^{p+1} \mathcal{A}$ as a linear operator, unambiguously by

$$\delta(a_0 \delta a_1 \cdots \delta a_p) =: \delta a_0 \delta a_1 \cdots \delta a_p \, . \qquad (7.15)$$

It is then easily seen to satisfy the basic relations

$$\delta^2 = 0 \, , \qquad (7.16)$$
$$\delta(\omega_1 \omega_2) = \delta(\omega_1) \omega_2 + (-1)^p \omega_1 \delta \omega_2 \, , \quad \omega_1 \in \Omega^p \mathcal{A} \, , \quad \omega_2 \in \Omega \mathcal{A} \, . (7.17)$$

Notice that there is nothing like graded commutativity of forms, that is there are no relationships of the type $\omega_{(p)} \omega_{(q)} = (-1)^{pq} \omega_{(q)} \omega_{(p)}$, with $\omega_{(i)} \in \Omega^i \mathcal{A}$.

The graded differential algebra $(\Omega \mathcal{A}, \delta)$ is characterized by the following universal property [30, 83],

Proposition 7.1.2. *Let (Γ, Δ) be a graded differential algebra, $\Gamma = \oplus_p \Gamma^p$, and let $\rho : A \to \Gamma^0$ be a morphism of unital algebras. Then, there exists an extension of ρ to a morphism of graded differential algebras $\tilde{\rho} : \Omega A \to \Gamma$, and this extension is unique,*

$$
\begin{array}{ccc}
\tilde{\rho}: & \Omega^p A & \longrightarrow \quad \Gamma^p \\[4pt]
\delta \downarrow & \quad \downarrow \Delta \quad , & \tilde{\rho} \circ \delta = \Delta \circ \tilde{\rho} . \qquad (7.18) \\[4pt]
\tilde{\rho}: & \Omega^{p+1} A & \longrightarrow \quad \Gamma^{p+1}
\end{array}
$$

Proof. Given the morphism $\rho : A \to \Gamma^0$, one defines $\tilde{\rho} : \Omega^p A \to \Gamma^p$ by

$$
\tilde{\rho}(a_0 \delta a_1 \cdots \delta a_p) =: \rho(a_0) \Delta(\rho(a_1)) \cdots \Delta(\rho(a_p)) . \qquad (7.19)
$$

This map is uniquely defined by ρ since $\Omega^p A$ is spanned as a left A-module by monomials $a_0 \delta a_1 \cdots \delta a_p$. Next, the identity (7.13) and its counterpart for the elements $\rho(a_i)$ and the derivation Δ ensure that products are sent to products. Finally, by using (7.15) and the fact that Δ is a derivation, one has

$$
\begin{aligned}
(\tilde{\rho} \circ \delta)(a_0 \delta a_1 \cdots \delta a_p) &= \tilde{\rho}(\delta a_0 \delta a_1 \cdots \delta a_p) \\
&= \Delta \rho(a_0) \Delta(\rho(a_1)) \cdots \Delta(\rho(a_p)) \\
&= \Delta((\rho(a_0)) \Delta(\rho(a_1)) \cdots \Delta(\rho(a_p)) \\
&= (\Delta \circ \tilde{\rho})(a_0 \delta a_1 \cdots \delta a_p) , \qquad (7.20)
\end{aligned}
$$

which proves the commutativity of the diagram (7.18): $\tilde{\rho} \circ \delta = \Delta \circ \tilde{\rho}$.

The universal algebra ΩA is not very interesting from the cohomological point of view. From the very definition of δ in (7.15), it follows that all cohomology spaces $H^p(\Omega A) =: Ker(\delta : \Omega^p A \to \Omega^{p+1} A)/Im(\delta : \Omega^{p-1} A \to \Omega^p A)$ vanish, except in degree zero where $H^0(\Omega A) = \mathbb{C}$.

We shall now explicitly construct the algebra ΩA in terms of tensor products. Firstly, consider the submodule of $A \otimes_{\mathbb{C}} A$ given by

$$
ker(m : A \otimes_{\mathbb{C}} A \to A) , \quad m(a \otimes_{\mathbb{C}} b) = ab . \qquad (7.21)
$$

This submodule is generated by elements of the form $1 \otimes_{\mathbb{C}} a - a \otimes_{\mathbb{C}} 1$ with $a \in A$. Indeed, if $\sum a_i b_i = m(\sum a_i \otimes_{\mathbb{C}} b_i) = 0$, then one gets $\sum a_i \otimes_{\mathbb{C}} b_i = \sum a_i (1 \otimes_{\mathbb{C}} b_i - b_i \otimes_{\mathbb{C}} 1)$. Furthermore, the map $\Delta : A \to ker(m : A \otimes_{\mathbb{C}} A \to A)$ defined by $\Delta a =: 1 \otimes_{\mathbb{C}} a - a \otimes_{\mathbb{C}} 1$, satisfies the analogue of (7.1), $\Delta(ab) = (\Delta a)b + a \Delta b$. There is an isomorphism of bimodules

$$
\begin{aligned}
\Omega^1 A &\simeq ker(m : A \otimes_{\mathbb{C}} A \to A) , \\
\delta a &\leftrightarrow 1 \otimes_{\mathbb{C}} a - a \otimes_{\mathbb{C}} 1 ,
\end{aligned}
$$

$$
\text{or} \qquad \sum a_i \delta b_i \leftrightarrow \sum a_i (1 \otimes_{\mathbb{C}} b_i - b_i \otimes_{\mathbb{C}} 1) . \qquad (7.22)
$$

By identifying $\Omega^1 \mathcal{A}$ with the space $ker(m : \mathcal{A} \otimes_{\mathbb{C}} \mathcal{A} \to \mathcal{A})$ the differential is given by

$$\delta : \mathcal{A} \to \Omega^1 \mathcal{A} , \quad \delta a = 1 \otimes_{\mathbb{C}} a - a \otimes_{\mathbb{C}} 1 . \tag{7.23}$$

As for forms of higher degree one has, then,

$$\Omega^p \mathcal{A} \simeq \underbrace{\Omega^1 \mathcal{A} \otimes_{\mathcal{A}} \cdots \otimes_{\mathcal{A}} \Omega^1 \mathcal{A}}_{p-times} \subset \underbrace{\mathcal{A} \otimes_{\mathbb{C}} \cdots \otimes_{\mathbb{C}} \mathcal{A}}_{(p+1)-times} ,$$

$$a_0 \delta a_1 \delta a_2 \cdots \delta a_p \mapsto a_0(1 \otimes_{\mathbb{C}} a_1 - a_1 \otimes_{\mathbb{C}} 1) \otimes_{\mathcal{A}} \cdots \otimes_{\mathcal{A}} (1 \otimes_{\mathbb{C}} a_p - a_p \otimes_{\mathbb{C}} 1) ,$$

$$\forall \, a_k \in \mathcal{A} . \tag{7.24}$$

For instance, the image of the two form $a_0 \delta a_1 \delta a_2$ is given by

$$a_0(1 \otimes_{\mathbb{C}} a_1 - a_1 \otimes_{\mathbb{C}} 1) \otimes_{\mathcal{A}} (1 \otimes_{\mathbb{C}} a_2 - a_2 \otimes_{\mathbb{C}} 1)$$
$$= a_0 \otimes_{\mathbb{C}} a_1 \otimes_{\mathbb{C}} a_2 - a_0 \otimes_{\mathbb{C}} a_1 a_2 \otimes_{\mathbb{C}} 1$$
$$-a_0 a_1 \otimes_{\mathbb{C}} 1 \otimes_{\mathbb{C}} a_2 + a_0 a_1 \otimes_{\mathbb{C}} a_2 \otimes_{\mathbb{C}} 1) \subset \mathcal{A} \otimes_{\mathbb{C}} \mathcal{A} \otimes_{\mathbb{C}} \mathcal{A} . \tag{7.25}$$

The multiplication and the bimodule structures are given by,

$$(\omega_1 \otimes_{\mathcal{A}} \cdots \otimes_{\mathcal{A}} \omega_p) \cdot (\omega_{p+1} \otimes_{\mathcal{A}} \cdots \otimes_{\mathcal{A}} \omega_{p+q}) =: \omega_1 \otimes_{\mathcal{A}} \cdots \otimes_{\mathcal{A}} \omega_{p+q} ,$$
$$a \cdot (\omega_1 \otimes_{\mathcal{A}} \cdots \otimes_{\mathcal{A}} \omega_p) =: (a\omega_1) \otimes_{\mathcal{A}} \cdots \otimes_{\mathcal{A}} \omega_p ,$$
$$(\omega_1 \otimes_{\mathcal{A}} \cdots \otimes_{\mathcal{A}} \omega_p) \cdot a =: \omega_1 \otimes_{\mathcal{A}} \cdots \otimes_{\mathcal{A}} (\omega_p a) ,$$
$$\forall \, \omega_j \in \Omega^1 \mathcal{A} , a \in \mathcal{A} . \tag{7.26}$$

The realization of the differential δ is also easily found. Firstly, consider any one-form $\omega = \sum a_i \otimes_{\mathbb{C}} b_i = \sum a_i(1 \otimes_{\mathbb{C}} b_i - b_i \otimes_{\mathbb{C}} 1)$ (since $\sum a_i b_i = 0$). Its differential $\delta\omega \in \Omega^1 \mathcal{A} \otimes_{\mathcal{A}} \Omega^1 \mathcal{A}$ is given by

$$\delta\omega \quad =: \quad \sum(1 \otimes_{\mathbb{C}} a_i - a_i \otimes_{\mathbb{C}} 1) \otimes_{\mathcal{A}} (1 \otimes_{\mathbb{C}} b_i - b_i \otimes_{\mathbb{C}} 1)$$
$$= \quad \sum 1 \otimes_{\mathbb{C}} a_i \otimes_{\mathbb{C}} b_i - a_i \otimes_{\mathbb{C}} 1 \otimes_{\mathbb{C}} b_i + a_i \otimes_{\mathbb{C}} b_i \otimes_{\mathbb{C}} 1 . \tag{7.27}$$

Then δ is extended by using the Leibniz rule with respect to the product $\otimes_{\mathcal{A}}$,

$$\delta(\omega_1 \otimes_{\mathcal{A}} \cdots \otimes_{\mathcal{A}} \omega_p) =: \sum_{i=1}^{p} (-1)^{i+1} \omega_1 \otimes_{\mathcal{A}} \cdots \otimes_{\mathcal{A}} \delta\omega_i \otimes_{\mathcal{A}} \cdots \otimes_{\mathcal{A}} \omega_p , \quad \forall \, \omega_j \in \Omega^1 \mathcal{A} . \tag{7.28}$$

Finally, we mention that if $\mathcal{A}$ has an involution $*$, the algebra $\Omega\mathcal{A}$ is also made into an involutive algebra by defining

$$(\delta a)^* \quad =: \quad -\delta a^* , \quad \forall \, a \in \mathcal{A} \tag{7.29}$$
$$(a_0 \delta a_1 \cdots \delta a_p)^* \quad =: \quad (\delta a_p)^* \cdots (\delta a_1)^* a_0^*$$
$$= \quad a_p^* \delta a_{p-1}^* \cdots \delta a_0^*$$
$$+ \sum_{i=0}^{p-1} (-1)^{p+i} \delta a_p^* \cdots \delta(a_{i+1}^* a_i^*) \cdots \delta a_0^* . \tag{7.30}$$

7.1.1 The Universal Algebra of Ordinary Functions

Take $\mathcal{A} = \mathcal{F}(M)$, with $\mathcal{F}(M)$ the algebra of complex valued, continuous functions on a topological space M, or of smooth functions on a manifold M (or some other algebra of functions). Then, identify (a suitable completion of) $\mathcal{A} \otimes_{\mathbb{C}} \cdots \otimes_{\mathbb{C}} \mathcal{A}$ with $\mathcal{F}(M \times \cdots \times M)$. If $f \in \mathcal{A}$, then

$$\delta f(x_1, x_2) =: (1 \otimes_{\mathbb{C}} f - f \otimes_{\mathbb{C}} 1)(x_1, x_2) = f(x_2) - f(x_1) . \tag{7.31}$$

Therefore, $\Omega^1 \mathcal{A}$ can be identified with the space of functions of two variables vanishing on the diagonal. In turn, $\Omega^p \mathcal{A}$ is identified with the set of functions f of $p + 1$ variables vanishing on contiguous diagonals,

$$f(x_1, \cdots, x_{k-1}, x, x, x_{k+2}, \cdots, x_{p+1}) = 0 . \tag{7.32}$$

The differential is given by,

$$\delta f(x_1, \cdots x_{p+1}) =: \sum_{k=1}^{p+1} (-1)^{k-1} f(x_1, \cdots, x_{k-1}, x_{k+1}, \cdots, x_{p+1}) . \tag{7.33}$$

The $\mathcal{A}$-bimodule structure is given by

$$\begin{aligned}
(gf)(x_1, \cdots x_{p+1}) &=: g(x_1) f(x_1, \cdots x_{p+1}) , \\
(fg)(x_1, \cdots x_{p+1}) &=: f(x_1, \cdots x_{p+1}) g(x_{p+1}) ,
\end{aligned} \tag{7.34}$$

and extends to the product of a p-form with a q-form as follows,

$$(fh)(x_1, \cdots x_{p+q}) =: f(x_1, \cdots x_{p+1}) h(x_{p+1}, \cdots x_{p+q}) , \tag{7.35}$$

Finally, the involution is simply given by

$$f^*(x_1, \cdots x_{p+1}) = (f(x_1, \cdots x_{p+1}))^* . \tag{7.36}$$

Notice that even if the algebra is commutative fh and hf are different with no relations among them (there is nothing like graded commutativity).

7.2 Connes' Differential Forms

Given a spectral triple $(\mathcal{A}, \mathcal{H}, D)$, one constructs an exterior algebra of forms by means of a suitable representation of the universal algebra $\Omega \mathcal{A}$ in the algebra of bounded operators on $\mathcal{H}$. The map

$$\begin{aligned}
\pi &: \Omega \mathcal{A} \longrightarrow \mathcal{B}(\mathcal{H}) , \\
\pi(a_0 \delta a_1 \cdots \delta a_p) &=: a_0 [D, a_1] \cdots [D, a_p] , \quad a_j \in \mathcal{A} ,
\end{aligned} \tag{7.37}$$

is clearly a homomorphism of algebras since both δ and $[D, \cdot]$ are derivations on $\mathcal{A}$. Furthermore, from $[D, a]^* = -[D, a^*]$, one finds that $\pi(\omega)^* = \pi(\omega^*)$ for any form $\omega \in \Omega \mathcal{A}$ and so π is a *-homomorphism.

One might think of defining the space of forms as the image $\pi(\Omega\mathcal{A})$. This is not possible, since in general, $\pi(\omega) = 0$ does not imply that $\pi(\delta\omega) = 0$. Such unpleasant forms ω, for which $\pi(\omega) = 0$ while $\pi(\delta\omega) \neq 0$, are called *junk forms*. They have to be disposed of in order to construct a true differential algebra and make π into a homomorphism of differential algebras.

Proposition 7.2.1. *Let* $J_0 =: \oplus_p J_0^p$ *be the graded two-sided ideal of* $\Omega\mathcal{A}$ *given by*

$$J_0^p =: \{\omega \in \Omega^p\mathcal{A}, \ \pi(\omega) = 0 \} \ . \tag{7.38}$$

Then, $J = J_0 + \delta J_0$ *is a graded differential two-sided ideal of* $\Omega\mathcal{A}$.

Proof. It is enough to show that J is a two-sided ideal, the property $\delta^2 = 0$ implying that it is differential. Take $\omega = \omega_1 + \delta\omega_2 \in J^p$, with $\omega_1 \in J_0^p$ and $\omega_2 \in J_0^{p-1}$. If $\eta \in \Omega^q\mathcal{A}$, then $\omega\eta = \omega_1\eta + (\delta\omega_2)\eta = \omega_1\eta + \delta(\omega_2\eta) - (-1)^{p-1}\omega_2\delta\eta = (\omega_1\eta - (-1)^{p-1}\omega_2\delta\eta) + \delta(\omega_2\eta) \in J^{p+q}$. Analogously, one finds that $\eta\omega \in J^{p+q}$.

Definition 7.2.1. *The graded differential algebra of Connes' forms over the algebra* $\mathcal{A}$ *is defined by*

$$\Omega_D\mathcal{A} =: \Omega\mathcal{A}/J \simeq \pi(\Omega\mathcal{A})/\pi(\delta J_0) \ . \tag{7.39}$$

It is naturally graded by the degrees of $\Omega\mathcal{A}$ and J, the space of p-forms being given by

$$\Omega_D^p\mathcal{A} = \Omega^p\mathcal{A}/J^p \ . \tag{7.40}$$

Since J is a differential ideal, the exterior differential δ defines a differential on $\Omega_D\mathcal{A}$,

$$d : \Omega_D^p\mathcal{A} \longrightarrow \Omega_D^{p+1}\mathcal{A} \ ,$$
$$d[\omega] =: [\delta\omega] \simeq [\pi(\delta\omega)] \ , \tag{7.41}$$

with $\omega \in \Omega^p\mathcal{A}$ and $[\omega]$ the corresponding class in $\Omega_D^p\mathcal{A}$.
Let us see the structure of the forms more explicitly.

- 0-forms.
  Since we take $\mathcal{A}$ to be a subalgebra of $\mathcal{B}(\mathcal{H})$, we have that $J \cap \Omega^0\mathcal{A} = J_0 \cap \mathcal{A} = \{0\}$. Thus

$$\Omega_D^0\mathcal{A} \simeq \mathcal{A} \ . \tag{7.42}$$

- 1-forms.
  We have $J \cap \Omega^1\mathcal{A} = J_0 \cap \Omega^1\mathcal{A} + J_0 \cap \Omega^0\mathcal{A} = J_0 \cap \Omega^1\mathcal{A}$. Thus,

$$\Omega_D^1\mathcal{A} \simeq \pi(\Omega^1\mathcal{A}) \ , \tag{7.43}$$

and this space coincides with the $\mathcal{A}$-bimodule of bounded operators on $\mathcal{H}$ of the form

$$\omega_1 = \sum_j a_0^j[D, a_1^j] \ , \quad a_i^j \in \mathcal{A} \ . \tag{7.44}$$

- 2-forms.

  We have $J \cap \Omega^2 \mathcal{A} = J_0 \cap \Omega^2 \mathcal{A} + J_0 \cap \Omega^1 \mathcal{A}$. Thus,

  $$\Omega_D^2 \mathcal{A} \simeq \pi(\Omega^2 \mathcal{A})/\pi(\delta(J_0 \cap \Omega^1 \mathcal{A})) . \tag{7.45}$$

  Therefore, the $\mathcal{A}$-bimodule $\Omega_D^2 \mathcal{A}$ of 2-forms is made up of classes of elements of the kind

  $$\omega_2 = \sum_j a_0^j [D, a_1^j][D, a_2^j] , \quad a_i^j \in \mathcal{A} , \tag{7.46}$$

  modulo the sub-bimodule of operators

  $$\left\{ \sum_j [D, b_0^j][D, b_1^j] \mid b_i^j \in \mathcal{A} , \sum_j b_0^j [D, b_1^j] = 0 \right\} . \tag{7.47}$$

- p-forms.

  In general, the $\mathcal{A}$-bimodule $\Omega_D^p \mathcal{A}$ of p-forms is given by

  $$\Omega_D^p \mathcal{A} \simeq \pi(\Omega^p \mathcal{A})/\pi(\delta(J_0 \cap \Omega^{p-1} \mathcal{A})) , \tag{7.48}$$

  and is made of classes of operators of the form

  $$\omega_p = \sum_j a_0^j [D, a_1^j][D, a_2^j] \cdots [D, a_p^j] , \quad a_i^j \in \mathcal{A} , \tag{7.49}$$

  modulo the sub-bimodule of operators

  $$\left\{ \sum_j [D, b_0^j][D, b_1^j] \cdots [D, b_{p-1}^j] \mid b_i^j \in \mathcal{A} , \sum_j b_0^j [D, b_1^j] \cdots [D, b_{p-1}^j] = 0 \right\} . \tag{7.50}$$

As for the exterior differential (7.41), it is given by

$$d \left[\sum_j a_0^j [D, a_1^j][D, a_p^j] \cdots [D, a_p^j] \right] = \left[\sum_j [D, a_0^j][D, a_1^j][D, a_2^j] \cdots [D, a_p^j] \right] . \tag{7.51}$$

7.2.1 The Usual Exterior Algebra

The methods described in the previous Section, when applied to the canonical triple over an ordinary manifold, reproduce the usual exterior algebra over that manifold. Consider the canonical triple $(\mathcal{A}, \mathcal{H}, D)$ on a closed n-dimensional Riemannian manifold M as described in Sect. 6.5. We recall that $\mathcal{A} = \mathcal{F}(M)$ is the algebra of smooth functions on M; $\mathcal{H} = L^2(M, S)$ is the Hilbert space of square integrable spinor fields over M; D is the Dirac operator of the Levi-Civita connection as given by (6.52). We immediately see that, for any $f \in \mathcal{A}$,

$$\pi(\delta f) =: [D, f] = \gamma^\mu(x)\partial_\mu f = \gamma(d_M f) , \qquad (7.52)$$

where $\gamma : \Gamma(M, C(M)) \longrightarrow \mathcal{B}(\mathcal{H})$ is the algebra morphism defined in (6.48) and d_M denotes the usual exterior derivative on M. In general, for $f_j \in \mathcal{A}$,

$$\pi(f_0 \delta f_1 \ldots \delta f_p) =: f_0[D, f_1] \ldots [D, f_p] = \gamma(f_0 d_M f_1 \cdot \ldots \cdot d_M f_p) , \qquad (7.53)$$

where the differentials $d_M f_j$ are now regarded as sections of the Clifford bundle $C_1(M)$ (while the functions f_j can be thought of as sections of $C_0(M)$) and the dot $\cdot$ denotes the Clifford product in the fibers of $C(M) = \oplus_k C_k(M)$. It is worth noticing that the image of the map π is made up of multiplicative operators on the Hilbert space $\mathcal{H}$.

Since a generic differential 1-form on M can be written as $\sum_j f_0^j d_M f_1^j$, with $f_0^j, f_1^j \in \mathcal{A}$, using (7.52) we can identify Connes' 1-forms $\Omega_D^1 \mathcal{A}$ with the usual differential 1-forms $\Lambda^1(M)$,

$$\Omega_D^1 \mathcal{A} \simeq \Lambda^1(M) . \qquad (7.54)$$

To be more precise, we are really identifying the space $\Omega_D^1 \mathcal{A}$ with the image in $\mathcal{B}(\mathcal{H})$, through the morphism γ, of the space $\Lambda^1(M)$.

Next, we analyze the junk 2-forms. For any $f \in \mathcal{A}$, consider the universal forms

$$\alpha = \frac{1}{2}(f \delta f - (\delta f)f) \neq 0 ,$$
$$\delta\alpha = \delta f \delta f . \qquad (7.55)$$

One easily finds that

$$\pi(\alpha) = \frac{1}{2}\gamma^\mu(f\partial_\mu f - (\partial_\mu f)f) = 0 ,$$
$$\pi(\delta\alpha) = \gamma^\mu\gamma^\nu\partial_\mu f\partial_\nu f$$
$$= \frac{1}{2}(\gamma^\mu\gamma^\nu + \gamma^\nu\gamma^\mu)\partial_\mu f\partial_\nu f + \frac{1}{2}(\gamma^\mu\gamma^\nu - \gamma^\nu\gamma^\mu)\partial_\mu f\partial_\nu f$$
$$= (-g^{\mu\nu}\partial_\mu f\partial_\nu f)\mathbb{I}_{2^{[n/2]}} \neq 0 . \qquad (7.56)$$

Here we have used (6.49), $g^{\mu\nu}$ being the components of the metric. We conclude that the form $\delta\alpha$ is a junk 2-form. A generic junk 2-form is a combination (with coefficients in $\mathcal{A}$) of forms like the one in (7.55). As a consequence, we infer from expression (7.56) that $\pi(\delta(J_0 \cap \Omega^1 \mathcal{A}))$ is generated as an $\mathcal{A}$-module by the matrix $\mathbb{I}_{2^{[n/2]}}$. On the other hand, if $f_1, f_2 \in \mathcal{A}$, from (7.53) we have that

$$\pi(\delta f_1 \delta f_2) = \gamma(d_M f_1 \cdot d_M f_2)$$
$$= \gamma^\mu\gamma^\nu\partial_\mu f_1\partial_\nu f_2$$
$$= \frac{1}{2}(\gamma^\mu\gamma^\nu - \gamma^\nu\gamma^\mu)\partial_\mu f_1\partial_\nu f_2 + \frac{1}{2}(\gamma^\mu\gamma^\nu + \gamma^\nu\gamma^\mu)\partial_\mu f\partial_\nu f$$
$$= \gamma(d_M f_1 \wedge d_M f_2) - g(d_M f_1, d_M f_2)\mathbb{I}_{2^{[n/2]}} . \qquad (7.57)$$

Therefore, since a generic differential 2-form on M can be written as a sum $\sum_j f_0^j d_M f_1^j \wedge d_M f_2^j$, with $f_0^j, f_1^j, f_2^j \in \mathcal{A}$, by using (7.56) and (7.57) to eliminate junk forms, we can identify Connes' 2-forms $\Omega_D^2 \mathcal{A}$ with the image through γ of the usual differential 2-forms $\Lambda^2(M)$,

$$\Omega_D^2 \mathcal{A} \simeq \Lambda^2(M) . \tag{7.58}$$

The previous identifications can be generalized and one can identify (through the map γ)

$$\Omega_D^p \mathcal{A} \simeq \Lambda^p(M) . \tag{7.59}$$

In particular, $\Omega_D^p \mathcal{A} = 0$ if $p > dim M$.

To establish the identification (7.59) we need some additional facts from Clifford bundle theory which we take from [10].

For each $m \in M$, the Clifford algebra $C_m(M)$ has a natural filtration, $C_m(M) = \bigcup C_m^{(p)}$, where $C_m^{(p)}$ is spanned by products $\xi_1 \cdot \xi_2 \cdot \ldots \cdot \xi_k$, with $k \leq p$ and $\xi_j \in T_m^* M$. There is a natural graded algebra

$$grC_m =: \sum_p gr_p C_m , \quad gr_p C_m = C_m^{(p)} / C_m^{(p-1)} , \tag{7.60}$$

with a natural projection, the *symbol map*,

$$\sigma_p : C_m^{(p)} \longrightarrow gr_p C_m . \tag{7.61}$$

The graded algebra (7.60) is canonically isomorphic to the complexified exterior algebra $\Lambda_{\mathbb{C}}(T_m^* M)$, the isomorphism being given by

$$\Lambda_{\mathbb{C}}^p(T_m^* M) \ni \xi_1 \wedge \xi_2 \wedge \ldots \wedge \xi_p \longrightarrow \sigma_p(\xi_1 \cdot \xi_2 \cdot \ldots \cdot \xi_p) \in gr_p C_m . \tag{7.62}$$

Now, the Clifford algebra C_m also has a natural $\mathbb{Z}_2$ grading given by the parity of the number of terms ξ_j, $\xi_j \in T_m^*$ in a typical product $\xi_1 \cdot \xi_2 \cdot \ldots \cdot \xi_k$. If C_m^p is the subspace of $C_m^{(p)}$ made up of elements with the same parity as p, then the kernel of the symbol map σ_p, when restricted to C_m^p coincides with C_m^{p-2} and we have the identification[1]

$$\Lambda_{\mathbb{C}}^p(T_m^* M) \simeq C_m^p / C_m^{p-2} . \tag{7.63}$$

By considering the union over all points $m \in M$, one constructs the corresponding bundles over M. We shall still denote by σ_p the symbol map

$$\sigma_p : \Gamma(C^{(p)}) \longrightarrow \Gamma(gr_p C) . \tag{7.64}$$

But to avoid confusion we shall indicate by s_p its restriction to $\Gamma(C^p)$. We stress that

$$ker s_p \simeq \Gamma(C^{p-2}) . \tag{7.65}$$

---

[1] By using the canonical inner product given by the trace in the spinor representation, $\Lambda_{\mathbb{C}}^p(T_m^* M)$ can be identified with the orthogonal complement of the space $C_m^{(p-1)}$ in $C_m^{(p)}$ or equivalently, with the orthogonal complement of the space C_m^{p-2} in the space C_m^p.

Proposition 7.2.2. *Let $(\mathcal{A}, \mathcal{H}, D)$ be the canonical triple over the manifold M. Then,*

$$\pi(\delta(J_0 \cap \Omega^{p-1}\mathcal{A})) = \gamma(ker s_p) , \quad p \geq 2 . \tag{7.66}$$

Proof. Consider the universal $(p-1)$-form $\omega = \frac{1}{2}(f_0 \delta f_0 - \delta f_0 f_0) \delta f_1 \ldots \delta f_{p-2}$. Then $\pi(\omega) = 0$ and $\pi(\delta\omega) = \gamma(-\|d_M f_0\|^2 d_M f_1 \cdot \ldots \cdot d_M f_{p-2})$. Since terms of the type $\|d_M f_0\|^2 d_M f_1 \cdot \ldots \cdot d_M f_{p-2}$ generate $\Gamma(C^{p-2})$ as an $\mathcal{A}$-module, one can find a universal form $\omega' \in \Omega^{p-1}\mathcal{A}$ with $\pi(\omega') = 0$ and $\pi(\delta\omega') = \gamma(\rho)$ where ρ is any given element of $\Gamma(C^{p-2})$. By using (7.65), this proves the inclusion $\gamma(ker s_p) \subseteq \pi(\delta(J_0 \cap \Omega^{p-1}\mathcal{A}))$.

Conversely, with $\omega = \sum_j f_0^j \delta f_1^j \ldots \delta f_{p-1}^j$ a generic universal $(p-1)$ form, the condition

$$\pi(\omega) =: \gamma^{\mu_1} \cdots \gamma^{\mu_{p-1}} \sum_j f_0^j \partial_{\mu_1} f_1^j \ldots \partial_{\mu_{p-1}} f_{p-1}^j = 0 , \tag{7.67}$$

implies, in particular, that

$$\sum_j f_0^j \partial_{[\mu_1} f_1^j \ldots \partial_{\mu_{p-1}]} f_{p-1}^j = 0 , \tag{7.68}$$

with the square brackets indicating complete anti-symmetrization of the enclosed indices. This condition readily yields the vanishing of the top component of

$$\pi(\delta\omega) =: \gamma^{\mu_0} \gamma^{\mu_1} \cdots \gamma^{\mu_{p-1}} \sum_j \partial_{\mu_0} f_0^j \partial_{\mu_1} f_1^j \ldots \partial_{\mu_{p-1}} f_{p-1}^j . \tag{7.69}$$

Indeed,

$$\sum_j \partial_{[\mu_0} f_0^j \partial_{\mu_1} f_1^j \ldots \partial_{\mu_{p-1}]} f_{p-1}^j = \partial_{[\mu_0} (\sum_j f_0^j \partial_{\mu_1} f_1^j \ldots \partial_{\mu_{p-1}]} f_{p-1}^j)$$

$$- \sum_j f_0^j \partial_{[\mu_0} \partial_{\mu_1} f_1^j \ldots \partial_{\mu_{p-1}]} f_{p-1}^j$$

$$= 0 . \tag{7.70}$$

It follows that $\pi(\delta\omega) \in \gamma(\Gamma(C^{p-2}) \simeq \gamma(ker s_p)$, and this proves the inclusion $\pi(\delta(J_0 \cap \Omega^{p-1}\mathcal{A})) \subseteq \gamma(ker s_p)$.

Proposition 7.2.3. *Let $(\mathcal{A}, \mathcal{H}, D)$ be the canonical triple over the manifold M. Then, a pair T_1 and T_2 of operators on the Hilbert space $\mathcal{H}$ is of the form $T_1 = \pi(\omega)$, $T_2 = \pi(\delta\omega)$ for some universal form $\omega \in \Omega^p A$, if and only if there are sections ρ_1 of C^p and ρ_2 of C^{p+1}, such that*

$$T_j = \gamma(\rho_j) , \quad j = 1, 2 ,$$

$$d_M \sigma_p(\rho_1) = \sigma_{p+1}(\rho_2) . \tag{7.71}$$

Proof. If $\omega = f_0 \delta f_1 \ldots \delta f_p$, the identities $T_1 = \pi(\omega) = \gamma(f_0 \delta f_1 \ldots \delta f_p)$ and $T_2 = \pi(\omega) = \gamma(\delta f_0 \delta f_1 \ldots \delta f_p)$ will imply that $\rho_1 = f_0 d_M f_1 \cdots \cdot d_M f_p$, $\rho_2 = d_M f_0 \cdot d_M f_1 \cdots \cdot d_M f_p$, and in turn $\sigma_p(\rho_1) = f_0 d_M f_1 \wedge \ldots \wedge d_M f_p$, $\sigma_{p+1}(\rho_2) = d_M f_0 \wedge d_M f_1 \wedge \ldots \wedge d_M f_p$, and finally $d_M \sigma_p(\rho_1) = \sigma_{p+1}(\rho_2)$.

Conversely, if $\rho_1 \in \Gamma(C^p)$ and $\rho_2 \in \Gamma(C^{p+1})$ are such that $d_M \sigma_p(\rho_1) = \sigma_{p+1}(\rho_2)$, then ρ_2 is determined by ρ_1 up to an ambiguity in $\Gamma(C^{p-1})$. Suppose first that $\rho_2 \in \Gamma(C^{p-1}) \simeq ker s_{p+1}$; we can then take $\rho_1 = 0$. So one needs a universal form $\omega \in \Omega^p \mathcal{A}$ such that $\pi(\omega) = 0, \pi(\delta \omega) = \gamma(\rho_2)$. The existence of ω follows from the previous Proposition (for $p+1$).

Furthermore, if ρ_2 has components only in $\Gamma(C^{p+1}) \ominus \Gamma(C^{p-1})$, we can take $\rho_1 = \sum_j f_0^j d_M f_1^j \cdots \cdot d_M f_p^j$, $\rho_2 = \sum_j d_M f_0^j \cdot d_M f_1^j \cdots \cdot d_M f_p^j$. Then, the universal $\omega \in \Omega^p \mathcal{A}$ is just $\omega = \sum_j f_0^j \delta f_1^j \cdots \cdot \delta f_p^j$.

Proposition 7.2.4. *Let $(\mathcal{A}, \mathcal{H}, D)$ be the canonical triple over the manifold M. Then, the symbol map s_p gives an isomorphism*

$$s_p : \Omega_D^p \mathcal{A} \longrightarrow \Gamma(\Lambda_{\mathbb{C}}^p T^* M) , \qquad (7.72)$$

which commutes with the differential.

Proof. Firstly, one identifies $\pi(\Omega^p \mathcal{A})$ with $\Gamma(C^p)$ through γ. Then, Proposition 7.2.2 shows that $\pi(\delta(J_0 \cap \Omega^{p-1} \mathcal{A})) = ker s_p$. The commutativity with the differential follows from Proposition 7.2.3. Finally, one observes that from the definition of the symbol map, if $\rho_j \in \Gamma(C^{p_j})$, $j = 1, 2$, then

$$s_{p_1+p_2}(\rho_1 \rho_2) = s_{p_1}(\rho_1) \wedge s_{p_2}(\rho_2) \in \Gamma(\Lambda_{\mathbb{C}}^{p_1+p_2} T^* M) . \qquad (7.73)$$

As a consequence, the symbol maps s_p combine to yield an isomorphism of graded differential algebras

$$s : \Omega_D(C^\infty(M)) \longrightarrow \Gamma(\Lambda_{\mathbb{C}} T^* M) , \qquad (7.74)$$

which is also an isomorphism of $C^\infty(M)$-modules.

7.2.2 The Two-Point Space Again

As a very simple example, we shall now construct Connes' exterior algebra on the two-point space $Y = \{1, 2\}$ with the 0-dimensional even spectral triple $(\mathcal{A}, \mathcal{H}, D)$ constructed in Sect. 6.8. We already know that the associated algebra $\mathcal{A}$ of continuous function is the direct sum $\mathcal{A} = \mathbb{C} \oplus \mathbb{C}$ and any element $f \in \mathcal{A}$ is a couple of complex numbers (f_1, f_2), with $f_i = f(i)$ the value of f at the point i.

As we saw in Sect. 7.1.1, the space $\Omega^1 \mathcal{A}$ of universal 1-forms can be identified with the space of functions on $Y \times Y$ which vanish on the diagonal. Since the complement of the diagonal in $Y \times Y$ is made of two points, namely the couples $(1, 2)$ and $(2, 1)$, the space $\Omega^1 \mathcal{A}$ is 2-dimensional and a basis is constructed as follows. Consider the function e defined by $e(1) = 1, e(2) = 0$;

clearly, $(1-e)(1) = 0, (1-e)(2) = 1$. A possible basis for the 1-forms is then given by

$$e\delta e , \quad (1-e)\delta(1-e) . \tag{7.75}$$

Their values are given by

$$(e\delta e)(1,2) = -1 , \quad ((1-e)\delta(1-e))(1,2) = 0$$
$$(e\delta e)(2,1) = 0 , \quad ((1-e)\delta(1-e))(2,1) = -1 . \tag{7.76}$$

Any universal 1-form $\alpha \in \Omega^1\mathcal{A}$ can be written as $\alpha = \lambda e\delta e + \mu(1-e)\delta(1-e)$, with $\lambda, \mu \in \mathbb{C}$. As for the differential, $\delta : \mathcal{A} \to \Omega^1\mathcal{A}$, it is essentially a finite difference operator. For any $f \in \mathcal{A}$ one finds that

$$\delta f = (f_1 - f_2)e\delta e - (f_1 - f_2)(1-e)\delta(1-e) = (f_1 - f_2)\delta e . \tag{7.77}$$

As for the space $\Omega^p\mathcal{A}$ of universal p-forms, it can be identified with the space of functions of $p+1$ variables which vanish on contiguous diagonals. Since there are only two possible strings giving nonvanishing results, namely $(1,2,1,2,\cdots)$ and $(2,1,2,1,\cdots)$ the space $\Omega^p\mathcal{A}$ is two dimensional as well and one possible basis is given by

$$e(\delta e)^p , \quad (1-e)(\delta(1-e))^p . \tag{7.78}$$

The values taken by the first basis element are

$$(e(\delta e)^p)(1,2,1,2,\cdots) = \pm 1 , \tag{7.79}$$
$$(e(\delta e)^p)(2,1,2,1,\cdots) = 0 ; \tag{7.80}$$

and in (7.79) the plus (minus) sign occurs if the number of contiguous couples $(1,2)$ is even (odd). As for the second basis element we have

$$((1-e)(\delta(1-e))^p)(1,2,1,2,\cdots) = 0 , \tag{7.81}$$
$$((1-e)(\delta(1-e))^p)(2,1,2,1,\cdots) = \pm 1 , \tag{7.82}$$

where, in (7.82), the plus (minus) sign occurs if the number of contiguous couples $(2,1)$ is even (odd).

We now move to Connes' forms. We recall that the finite dimensional Hilbert space $\mathcal{H}$ is a direct sum $\mathcal{H} = \mathcal{H}_1 \oplus \mathcal{H}_2$; elements of $\mathcal{A}$ act as diagonal matrices $\mathcal{A} \ni f \mapsto \mathrm{diag}(f_1 \mathbb{I}_{dimH_1}, f_2 \mathbb{I}_{dimH_2})$; and D is an off diagonal operator $\begin{bmatrix} 0 & M^* \\ M & 0 \end{bmatrix}, M \in Lin(\mathcal{H}_1, \mathcal{H}_2)$.

One immediately finds that

$$\pi(e\delta e) =: e[D, e] = \begin{bmatrix} 0 & -M^* \\ 0 & 0 \end{bmatrix} ,$$

$$\pi((1-e)\delta(1-e)) =: (1-e)[D, 1-e] = \begin{bmatrix} 0 & 0 \\ -M & 0 \end{bmatrix} , \tag{7.83}$$

and the representation of a generic 1-form $\alpha = \lambda e\delta e + \mu(1-e)\delta(1-e)$ is given by

$$\pi(\alpha) = -\begin{bmatrix} 0 & \lambda M^* \\ \mu M & 0 \end{bmatrix}. \tag{7.84}$$

As for the representation of 2-forms one gets

$$\pi(e\delta e\delta e) =: e[D,e][D,e] = \begin{bmatrix} -M^*M & 0 \\ 0 & 0 \end{bmatrix},$$

$$\pi((1-e)\delta(1-e)\delta(1-e)) =: (1-e)[D,1-e][D,1-e]$$

$$= \begin{bmatrix} 0 & 0 \\ 0 & -MM^* \end{bmatrix}. \tag{7.85}$$

In particular the operator $\pi(\delta\alpha)$ is readily found to be

$$\pi(\delta\alpha) = -(\lambda+\mu)\begin{bmatrix} M^*M & 0 \\ 0 & MM^* \end{bmatrix}, \tag{7.86}$$

from which we infer that there are no junk 1-forms. In fact, there are no junk forms whatsoever. Even forms are represented by diagonal operators while odd forms are represented by off diagonal ones.

7.3 Scalar Product for Forms

In order to define a scalar product for forms, we need another result which has been proven in [28]

Proposition 7.3.1. *Let $(\mathcal{A}, \mathcal{H}, D)$ be an n-dimensional spectral triple. Then, the state ϕ on $\mathcal{A}$ defined by*

$$\phi(a) =: tr_\omega(\pi(a)|D|^{-n}), \quad \forall\, a \in \mathcal{A}, \tag{7.87}$$

with tr_ω denoting the Dixmier trace, is a trace state, i.e.

$$\phi(ab) = \phi(ba), \quad \forall\, a \in \mathcal{A}. \tag{7.88}$$

Thus, we get a positive trace on $\mathcal{A}$ as was alluded to at the end of Sect. 6.6.

In fact, one needs to extend the previous result to all of $\pi(\Omega\mathcal{A})$. Now, this is not possible in general and (rather mild, indeed) regularity conditions on the algebra are required. Let us recall from Sect. 6.4, that the subalgebra $\mathcal{A}^2$ of $\mathcal{A}$ was generated by elements $a \in \mathcal{A}$ such that both a and $[D,a]$ are in the domain of the derivation δ defined in (6.41). In [28] it is proven that the state ϕ on $\pi(\Omega\mathcal{A})$ defined by

$$\phi(T) =: tr_\omega(T|D|^{-n}), \quad \forall\, T \in \pi(\Omega\mathcal{A}), \tag{7.89}$$

is a trace state,

$$\phi(ST) = \phi(TS) \, , \quad \forall \, S, T \in \pi(\Omega \mathcal{A}) \, , \tag{7.90}$$

provided that $\mathcal{A}^2 = \mathcal{A}$ (or that $\mathcal{A}^2$ is a large enough subalgebra). We refer to [28] for the proofs of the previous statements. Here we only remark that, by using the cyclic property of tr_ω, the condition (7.90) is equivalent to one of the following,

$$\begin{aligned} tr_\omega([T, |D|^{-n}]) &= 0 \, , && \forall \, T \in \pi(\Omega \mathcal{A}) \, , \\ tr_\omega(ST|D|^{-n}) &= tr_\omega(S|D|^{-n}T) \, , && \forall \, S, T \in \pi(\Omega \mathcal{A}) \, . \end{aligned} \tag{7.91}$$

Property (7.90) (together with the cyclic property of tr_ω) implies that the following three traces coincide and can be taken as a definition of an inner product on $\pi(\Omega^p \mathcal{A})$,[2]

$$\begin{aligned} \langle T_1, T_2 \rangle_p \; &=: \; tr_\omega(T_1^* T_2 |D|^n) \\ &= \; tr_\omega(T_1^* |D|^n T_2) \\ &= \; tr_\omega(T_2 |D|^n T_1^*) \, , \quad \forall \, T_1, T_2 \in \pi(\Omega^p \mathcal{A}) \, . \end{aligned} \tag{7.92}$$

Forms of different degree are defined to be orthogonal.
Now let $\widetilde{\mathcal{H}}_p$ be the corresponding completion of $\pi(\Omega^p \mathcal{A})$. With $a \in \mathcal{A}$ and $T_1, T_2 \in \pi(\Omega^p \mathcal{A})$, we obtain

$$\langle aT_1, aT_2 \rangle_p = tr_\omega(T_1^* a^* |D|^n a T_2) = tr_\omega(T_2 |D|^n T_1^* a^* a) \, , \tag{7.93}$$
$$\langle T_1 a, T_2 a \rangle_p = tr_\omega(a^* T_1^* |D|^n T_2 a) = tr_\omega(a^* a T_1^* |D|^n T_2) \, . \tag{7.94}$$

As a consequence, the unitary group $\mathcal{U}(\mathcal{A})$ of $\mathcal{A}$,

$$\mathcal{U}(\mathcal{A}) =: \{ u \in \mathcal{A} \mid u^* u = uu^* = 1 \} \, , \tag{7.95}$$

has two commuting unitary representations, L and R, on $\widetilde{\mathcal{H}}_p$ given by left and right multiplications. Now, as $\pi(\delta(J_0 \cap \Omega^{p-1} \mathcal{A})$ is a submodule of $\pi(\Omega^p \mathcal{A})$, its closure in $\widetilde{\mathcal{H}}_p$ is left invariant by these two representations. Let P_p be the orthogonal projection of $\widetilde{\mathcal{H}}_p$, with respect to the inner product (7.92), which projects onto the orthogonal complement of $\pi(\delta(J_0 \cap \Omega^{p-1} \mathcal{A}))$. Then P_p commutes with $L(a)$ and $R(a)$, if $a \in \mathcal{U}(\mathcal{A})$ and so for any $a \in \mathcal{A}$. Define $\mathcal{H}_p = P_p \widetilde{\mathcal{H}}_p$; this space also coincides with the completion of the Connes' forms $\Omega_D^p \mathcal{A}$. The left and right representations of $\mathcal{A}$ on $\widetilde{\mathcal{H}}_p$ reduce to algebra representations on $\mathcal{H}_p$ which extend the left and right module action of $\mathcal{A}$ on $\Omega_D^p \mathcal{A}$.

---

[2] An alternative definition of the integral and of the inner product for forms which is based on the heat kernel expansion and uses the heat operator $exp(-\varepsilon D^2)$ has been devised in [67]. When applied to the canonical triple over a manifold it gives the usual results.

As an example, consider again the algebra $\mathcal{A} = C^\infty(M)$ and the associated canonical triple $(\mathcal{A}, \mathcal{H}, D)$ over a manifold M of dimension $n = \dim M$. Then, the trace requirement (7.90) is satisfied.[3] Furthermore,

Proposition 7.3.2. *With the canonical isomorphism between $\Omega_D\mathcal{A}$ and $\Gamma(\Lambda_{\mathbb{C}}T^*M)$ described in Sec. 7.2.1, the inner product on $\Omega_D^p\mathcal{A}$ is proportional to the Riemannian inner product on p-forms,*

$$\langle \omega_1, \omega_2 \rangle_p = (-1)^p \frac{2^{[n/2]+1-n}\pi^{-n/2}}{n\Gamma(n/2)} \int_M \omega_1 \wedge^* \omega_2 ,$$

$$\forall\ \omega_1, \omega_2 \in \Omega_D^p\mathcal{A} \simeq \Gamma(\Lambda_{\mathbb{C}}T^*M) . \tag{7.96}$$

Proof. If $T \in \Omega^p\mathcal{A}$ and $\rho \in \Gamma(C^p)$, with $\pi(T) = \gamma(\rho)$, we have that $P_p\pi(T) = \gamma(\omega) \in \mathcal{H}_p$, with ω the component of ρ in $\Gamma(C^p \ominus C^{p-1})$. Using the trace theorem 6.3.1, we get

$$\begin{aligned}
\langle \gamma(\omega_1), \gamma(\omega_2) \rangle_p &= tr_\omega(\omega_1^*\omega_2 |D|^{-n}) \\
&= \frac{1}{n(2\pi)^n} \int_{S^*M} tr\sigma_{-n}(\gamma(\omega_1)^*\gamma(\omega_2)|D|^{-n}) \\
&= \frac{1}{n(2\pi)^n} \left(\int_{S^{n-1}} d\xi \right) \int_M tr(\gamma(\omega_1)^*\gamma(\omega_2)dx \\
&= \frac{2^{1-n}\pi^{-n/2}}{n\Gamma(n/2)} \int_M tr(\gamma(\omega_1)^*\gamma(\omega_2))dx \\
&= (-1)^p \frac{2^{[n/2]+1-n}\pi^{-n/2}}{n\Gamma(n/2)} \int_M \omega_1 \wedge^* \omega_2 .
\end{aligned}$$

The last equality follows from the explicit (partially normalized) trace in the spin representation. Indeed,

$$\omega_j = \tfrac{1}{p!}\omega_{\mu_1\cdots\mu_p}^{(j)} dx^{\mu_1} \wedge \cdots \wedge dx^{\mu_p}, \quad j = 1, 2 , \quad \Rightarrow$$

$$\gamma(\omega_j) = \tfrac{1}{p!}\omega_{\mu_1\cdots\mu_p}^{(j)} \gamma^{\mu_1} \wedge \cdots \wedge \gamma^{\mu_p} = \tfrac{1}{p!}\omega_{\mu_1\cdots\mu_p}^{(j)} e_{a_1}^{\mu_1} \cdots e_{a_p}^{\mu_p} \gamma^{a_1} \wedge \cdots \wedge \gamma^{a_p} , \quad \Rightarrow$$

$$tr(\gamma(\omega_1)^*\gamma(\omega_2)) = (-1)^p 2^{[\frac{n}{2}]}\omega_{\mu_1\cdots\mu_p}^{(1)*} \omega_{\nu_1\cdots\nu_p}^{(2)} e_{a_1}^{\mu_1} \cdots e_{a_p}^{\mu_p} e_{b_1}^{\nu_1} \cdots e_{b_p}^{\nu_p} \eta^{a_1b_1} \cdots \eta^{a_pb_p}$$

$$= (-1)^p 2^{[\frac{n}{2}]}\omega_{\mu_1\cdots\mu_p}^{(1)*} \omega_{\nu_1\cdots\nu_p}^{(2)} g^{\mu_1\nu_1} \cdots g^{\mu_p\nu_p} , \tag{7.97}$$

from which one finds $tr(\gamma(\omega_1)^*\gamma(\omega_2))dx = (-1)^p 2^{[n/2]}\omega_1 \wedge^* \omega_2 .$

---

[3] In fact, in the commutative situation, the regularity condition does not play a crucial rôle. On a manifold of dimension n, the pseudo-differential operator $[T, |D|^{-n}]$, with T the image of a section of the Clifford bundle, is of order $n-1$ and its Dixmier trace vanishes (see Sect. 6.3). As a consequence, the first of (7.91) and then (7.90) are automatically satisfied.

8. Connections on Modules

As an example of the general situation, we shall start by describing the analogue of 'electromagnetism', namely the algebraic theory of connections (vector potentials) on a rank one trivial bundle (with fixed trivialization).

8.1 Abelian Gauge Connections

Suppose we are given a spectral triple $(\mathcal{A}, \mathcal{H}, D)$ from which we construct the algebra $\Omega_D \mathcal{A} = \oplus_p \Omega_D^p \mathcal{A}$ of forms. We also take it to be of dimension n.

Definition 8.1.1. *A vector potential V is a self-adjoint element of $\Omega_D^1 \mathcal{A}$. The corresponding field strength is the two-form $\theta \in \Omega_D^2 \mathcal{A}$ defined by*

$$\theta = dV + V^2 . \tag{8.1}$$

Thus, V is of the form $V = \sum_j a_j [D, b_j]$, $a_j, b_j \in \mathcal{A}$ with V self-adjoint, $V^* = V$. Notice that, although V can be written in several ways as a sum, its exterior derivative $dV \in \Omega_D^2 \mathcal{A}$ is defined unambiguously. It can though be written in several ways as a sum, $dV = \sum_j [D, a_j][D, b_j]$, modulo junk. The curvature θ is self-adjoint as well. It is evident that V^2 is self-adjoint if V is. As for dV, we have,

$$dV - (dV)^* = \sum_j [D, a_j][D, b_j] - \sum_j [D, b_j^*][D, a_j^*] . \tag{8.2}$$

Since $V^* = -\sum_j [D, b_j^*] a_j^* = -\sum_j [D, b_j^* a_j^*] + \sum_j b_j^* [D, a_j^*]$ and $V - V^* = 0$, we get that the following is a junk 2-form,

$$j_2 = dV - dV^* = \sum_j [D, a_j][D, b_j] - \sum_j [D, b_j^*][D, a_j^*] . \tag{8.3}$$

But j_2 is just the right-hand side of (8.2), and we infer that, modulo junk forms, $dV = (dV)^*$.

Since the algebra $\mathcal{A}$ is taken to be a unital *-algebra, it makes sense to consider the group $\mathcal{U}(\mathcal{A})$ of unitary elements of $\mathcal{A}$,

$$\mathcal{U}(\mathcal{A}) =: \{ u \in \mathcal{A} \mid uu^* = u^*u = \mathbb{I} \} . \tag{8.4}$$

This group provides the (infinite dimensional group of) gauge transformations.

Definition 8.1.2. *The unitary group $\mathcal{U}(\mathcal{A})$ acts on the vector potential V with the usual affine action*

$$(V, u) \longrightarrow V^u =: uVu^* + u[D, u^*] , \quad u \in \mathcal{U}(\mathcal{A}) . \tag{8.5}$$

The field strength θ will then transform with the adjoint action,

$$
\begin{aligned}
\theta^u &= dV^u + (V^u)^2 \\
&= duVu^* + udVu^* - uVdu^* + du[D, u^*] + uV^2u^* + \\
&\quad + uV[D, u^*] + u[D, u^*]uVu^* + u[D, u^*]u[D, u^*] \\
&= u(dV + V^2)u^* ,
\end{aligned}
\tag{8.6}
$$

where we have used the relations $du = [D, u^*]$ and $udu^* + (du)u^*$; the latter follows from $u^*u = \mathbb{I}$. Thus,

$$(\theta, u) \longrightarrow \theta^u = u\theta u^* , \quad u \in \mathcal{U}(\mathcal{A}) . \tag{8.7}$$

We can now introduce the analogue of the Yang-Mills functional.

Proposition 8.1.1. *1.   The functional*

$$YM(V) =: \langle dV + V^2, dV + V^2 \rangle_2 , \tag{8.8}$$

is positive, quartic and invariant under gauge transformations

$$V \longrightarrow V^u =: uVu^* + u[D, u^*] , \quad u \in \mathcal{U}(\mathcal{A}) . \tag{8.9}$$

2.   The functional

$$I(\alpha) =: tr_\omega(\pi(\delta\alpha + \alpha^2))^2|D|^{-n}) , \tag{8.10}$$

is positive, quartic and invariant on the space $\{\alpha \in \Omega^1\mathcal{A} \mid \alpha = \alpha^*\}$, under gauge transformations

$$\alpha \longrightarrow \alpha^u =: u\alpha u^* + u\delta u^* , \quad u \in \mathcal{U}(\mathcal{A}) . \tag{8.11}$$

3.

$$YM(V) = inf \{I(\alpha) \mid \pi(\alpha) = V\} . \tag{8.12}$$

Proof. Statements 1. and 2. are consequences of properties of the Dixmier trace and of the fact that both $dV + V^2$ and $\delta\alpha + \alpha^2$ transform 'covariantly' under gauge transformations. As for statement 3., it follows from the nearest-point property of an orthogonal projector: as an element of $\mathcal{H}_2$, $dV + V^2$ is equal to $P(\pi(\delta\alpha + \alpha^2))$ for any $\alpha \in \Omega^1\mathcal{A}$ such that $\pi(\alpha) = V$. Since the ambiguity in $\pi(\delta\alpha)$ is exactly $\pi(\delta(J_0 \cap \Omega^1\mathcal{A}))$, one has established 3.

Point 3. of Prop. 8.1.1 just states that the ambiguity in the definition of the curvature $\theta = dV + V^2$ can be ignored by taking the infimum $YM(V) = Inf \{tr_\omega\theta^2|D|^{-n}\}$ over all possibilities for $\theta = dV + V^2$, the exterior derivative $dV = \sum_j[D, a_0^j][D, a_1^j]$ being ambiguous.

As already mentioned, the module $\mathcal{E} = \mathcal{A}$ is just the analogue of a rank one trivial bundle with fixed trivialization so that one can identify the module of sections of the bundle with the complex-valued functions on the base.

8.1.1 Usual Electromagnetism

For the canonical triple $(\mathcal{A}, \mathcal{H}, D)$ over the manifold M, consider a 1-form $V \in \Lambda^1(M)$ and a universal 1-form $\alpha \in \Omega^1 \mathcal{A}$ such that $\sigma_1(\pi(\alpha)) = V$. Then $\sigma_2(\pi(\delta\alpha)) = d_M V$. From proposition 7.2.3, for any two such α's, the corresponding operators $\pi(\delta\alpha)$ differ by an element of $\pi(\delta(J_0 \cap \Omega^1 \mathcal{A}) = ker\sigma_2$. Then, by using (7.96)

$$
\begin{aligned}
YM(V) &= inf \{I(\alpha) \mid \pi(\alpha) = V\} = \langle d_M V, d_M V \rangle_2 \\
&= \frac{2^{[n/2]+1-n}\pi^{-n}}{n\Gamma(n/2)} \int \|d_M V\|^2 dx ,
\end{aligned}
\tag{8.13}
$$

which is (proportional to) the usual abelian gauge action.

8.2 Universal Connections

We now introduce the notion of a connection on a (finite projective) module. We shall do it with respect to the universal calculus $\Omega\mathcal{A}$ introduced in Sect. 7.1 as this is the prototype for any calculus. So, to be precise, by a connection we really mean an universal connection although we drop the adjective universal whenever there is no risk of confusion.

Definition 8.2.1. *A (universal) connection on the right $\mathcal{A}$-module $\mathcal{E}$ is a $\mathbb{C}$-linear map*

$$
\nabla : \mathcal{E} \otimes_\mathcal{A} \Omega^p \mathcal{A} \longrightarrow \mathcal{E} \otimes_\mathcal{A} \Omega^{p+1} \mathcal{A} ,
\tag{8.14}
$$

defined for any $p \geq 0$, and satisfying the Leibniz rule

$$
\nabla(\omega\rho) = (\nabla\omega)\rho + (-1)^p \omega\delta\rho , \quad \forall \omega \in \mathcal{E} \otimes_\mathcal{A} \Omega^p \mathcal{A} , \; \rho \in \Omega\mathcal{A} .
\tag{8.15}
$$

In this definition, the adjective universal refers to the use of the universal forms and to the fact that a connection constructed for any calculus can be obtained from a universal one via a projection much in the same way as any calculus can be obtained from the universal one. In Proposition 9.1.1 we shall explicitly construct the projection for the Connes' calculus.

A connection is completely determined by its restriction $\nabla : \mathcal{E} \to \mathcal{E} \otimes_\mathcal{A} \Omega^1 \mathcal{A}$, which satisfies

$$
\nabla(\eta a) = (\nabla\eta)a + \eta \otimes_\mathcal{A} \delta a , \quad \forall \eta \in \mathcal{E} , \; a \in \mathcal{A} .
\tag{8.16}
$$

This is then extended by using the Leibniz rule (8.15).

Proposition 8.2.1. *The composition,*

$$
\nabla^2 = \nabla \circ \nabla : \mathcal{E} \otimes_\mathcal{A} \Omega^p \mathcal{A} \longrightarrow \mathcal{E} \otimes_\mathcal{A} \Omega^{p+2} \mathcal{A} ,
\tag{8.17}
$$

is $\Omega\mathcal{A}$-linear.

Proof. By condition (8.15) one has

$$\begin{aligned}
\nabla^2(\omega\rho) &= \nabla\left((\nabla\omega)\rho + (-1)^p\omega\delta\rho\right) \\
&= (\nabla^2\omega)\rho + (-1)^{p+1}(\nabla\omega)\delta\rho + (-1)^p(\nabla\omega)\delta\rho + \omega\delta^2\rho \\
&= (\nabla^2\omega)\rho \ .
\end{aligned} \tag{8.18}$$

The restriction of ∇^2 to $\mathcal{E}$ is the *curvature*

$$\theta : \mathcal{E} \to \mathcal{E} \otimes_A \Omega^2 A \ , \tag{8.19}$$

of the connection. By (8.15) it is A-linear, $\theta(\eta a) = \theta(\eta)a$ for any $\eta \in \mathcal{E}, a \in A$, and satisfies

$$\nabla^2(\eta \otimes_A \rho) = \theta(\eta)\rho \ , \quad \forall \eta \in \mathcal{E} \ , \ \rho \in \Omega A \ . \tag{8.20}$$

Since $\mathcal{E}$ is projective, any A-linear map : $\mathcal{E} \to \mathcal{E} \otimes_A \Omega A$ can be thought of as a matrix with entries in ΩA or as an element in $End_A\mathcal{E} \otimes_A \Omega A$. In particular, the curvature θ can be thought of as an element of $End_A\mathcal{E} \otimes_A \Omega^2 A$. Furthermore, by viewing any element of $End_A\mathcal{E} \otimes_A \Omega A$ as a map from $\mathcal{E}$ into $\mathcal{E} \otimes_A \Omega A$, the connection ∇ on $\mathcal{E}$ determines a connection $[\nabla, \cdot]$ on $End_A\mathcal{E}$ by

$$[\nabla, \cdot] : End_A\mathcal{E} \otimes_A \Omega^p A \longrightarrow End_A\mathcal{E} \otimes_A \Omega^{p+1} A \ ,$$
$$[\nabla, \alpha] =: \nabla \circ \alpha - \alpha \circ \nabla \ , \quad \forall \alpha \in End_A\mathcal{E} \otimes_A \Omega^p A. \tag{8.21}$$

Proposition 8.2.2. *The curvature θ satisfies the following Bianchi identity,*

$$[\nabla, \theta] = 0 \ . \tag{8.22}$$

Proof. Since $\theta : \mathcal{E} \to \Omega^2 A$, the map $[\nabla, \theta]$ makes sense. Furthermore,

$$[\nabla, \theta] = \nabla \circ \nabla^2 - \nabla^2 \circ \nabla = \nabla^3 - \nabla^3 = 0 \ . \tag{8.23}$$

Connections always exist on a projective module. To start with, let us consider the case of a free module $\mathcal{E} = \mathbb{C}^N \otimes_{\mathbb{C}} A \simeq A^N$. Forms with values in $\mathbb{C}^N \otimes_{\mathbb{C}} A$ can be identified canonically with

$$\mathbb{C}^N \otimes_{\mathbb{C}} \Omega A = (\mathbb{C}^N \otimes_{\mathbb{C}} A) \otimes_A \Omega A \simeq (\Omega A)^N \ . \tag{8.24}$$

Then, a connection is given by the operator

$$\nabla_0 = \mathbb{I} \otimes \delta : \mathbb{C}^N \otimes_{\mathbb{C}} \Omega^p A \longrightarrow \mathbb{C}^N \otimes_{\mathbb{C}} \Omega^{p+1} A \ . \tag{8.25}$$

If we think of ∇_0 as acting on $(\Omega A)^N$ we can represent it as the operator $\nabla_0 = (\delta, \delta, \cdots, \delta)$ (N-times).
Consider a generic projective module $\mathcal{E}$, and let $p : \mathbb{C}^N \otimes_{\mathbb{C}} A \to \mathcal{E}$ and $\lambda : \mathcal{E} \to \mathbb{C}^N \otimes_{\mathbb{C}} A$ be the corresponding projection and inclusion maps as in Sect. 4.2. On $\mathcal{E}$ there is a connection ∇_0 given by the composition

$$\mathcal{E} \otimes_A \Omega^p A \xrightarrow{\lambda} \mathbb{C}^N \otimes_{\mathbb{C}} \Omega^p A \xrightarrow{\mathbb{I} \otimes \delta} \mathbb{C}^N \otimes_{\mathbb{C}} \Omega^{p+1} A \xrightarrow{p} \mathcal{E} \otimes_A \Omega^{p+1} A, \tag{8.26}$$

where we use the same symbol to denote the natural extension of the maps λ and p to $\mathcal{E}$-valued forms. The connection defined in (8.26) is called the *Grassmann connection* and is explicitly given by

$$\nabla_0 = p \circ (\mathbb{I} \otimes \delta) \circ \lambda . \tag{8.27}$$

In what follows, we shall simply indicate it by

$$\nabla_0 = p\delta. \tag{8.28}$$

In fact, it turns out that the existence of a connection on the module $\mathcal{E}$ is completely equivalent to its being projective [40].

Proposition 8.2.3. *A right module has a connection if and only if it is projective.*

Proof. Consider the exact sequence of right $\mathcal{A}$-modules

$$0 \longrightarrow \mathcal{E} \otimes_{\mathcal{A}} \Omega^1 \mathcal{A} \overset{j}{\longrightarrow} \mathcal{E} \otimes_{\mathbb{C}} \mathcal{A} \overset{m}{\longrightarrow} \mathcal{E} \longrightarrow 0 , \tag{8.29}$$

where $j(\eta \delta a) = \eta \otimes a - \eta a \otimes 1$ and $m(\eta \otimes a) = \eta a$; both of these maps are (right) $\mathcal{A}$-linear. Now, as a sequence of vector spaces, (8.29) admits a splitting given by the section $s_0(\eta) = \eta \otimes 1$ of m, $m \circ s_0 = id_{\mathcal{E}}$. Furthermore, all such splittings form an affine space which is modeled over the space of linear maps from the base space $\mathcal{E}$ to the subspace $j(\mathcal{E} \otimes_{\mathcal{A}} \Omega^1 \mathcal{A})$. This means that there is a one to one correspondence between linear sections $s : \mathcal{E} \to \mathcal{E} \otimes_{\mathbb{C}} \mathcal{A}$ of m ($m \circ s = id_{\mathcal{E}}$) and linear maps $\nabla : \mathcal{E} \to \mathcal{E} \otimes_{\mathcal{A}} \Omega^1 \mathcal{A}$ given by

$$s = s_0 + j \circ \nabla , \quad s(\eta) = \eta \otimes 1 + j(\nabla \eta) , \quad \forall \, \eta \in \mathcal{E} . \tag{8.30}$$

Since

$$\begin{aligned} s(\eta a) - s(\eta)a &= \eta a \otimes 1 - \eta \otimes a + j(\nabla(\eta a)) - j(\nabla(\eta))a \\ &= j(\nabla(\eta a) - \nabla(\eta)a - \eta \delta a) , \end{aligned} \tag{8.31}$$

and as j is injective, we see that ∇ is a connection if and only if s is a right $\mathcal{A}$-module map,

$$\nabla(\eta a) - \nabla(\eta)a - \eta \delta a = 0 \quad \Leftrightarrow \quad s(\eta a) - s(\eta)a = 0 . \tag{8.32}$$

But such module maps exist if and only if $\mathcal{E}$ is projective: any right module map $s : \mathcal{E} \to \mathcal{E} \otimes_{\mathbb{C}} \mathcal{A}$ such that $m \circ s = \mathbb{I}_{\mathcal{E}}$ identifies $\mathcal{E}$ with a direct summand of the free module $\mathcal{E} \otimes_{\mathbb{C}} \mathcal{A}$, where the corresponding idempotent is $p = s \circ m$.

The previous proposition also says that the space $CC(\mathcal{E})$ of all universal connections on $\mathcal{E}$ is an affine space modeled on $End_{\mathcal{A}}\mathcal{E} \otimes_{\mathcal{A}} \Omega^1 \mathcal{A}$. Indeed, if ∇_1, ∇_2 are two connections on $\mathcal{E}$, their difference is $\mathcal{A}$-linear,

$$(\nabla_1 - \nabla_2)(\eta a) = ((\nabla_1 - \nabla_2)(\eta))a , \quad \forall \, \eta \in \mathcal{E} , \, a \in \mathcal{A} , \tag{8.33}$$

so that $\nabla_1 - \nabla_2 \in End_A \otimes_A \Omega^1 A$. By using (8.28) and (4.28) any connection can be written as

$$\nabla = p\delta + \alpha , \tag{8.34}$$

where α is any element in $End_A \mathcal{E} \otimes_A \Omega^1 A \simeq \mathbb{M}_N(\mathcal{A}) \otimes_A \Omega^1 A$ such that $\alpha = \alpha p = p\alpha = p\alpha p$; here p is the idempotent which identifies $\mathcal{E}$ as $\mathcal{E} = p\mathcal{A}^N$. The matrix of 1-forms α as in (8.34) is called the *gauge potential* of the connection ∇. For the corresponding curvature θ of ∇ we have

$$\theta = p\delta\alpha + \alpha^2 + p\delta p\delta p . \tag{8.35}$$

Indeed,

$$\begin{aligned}
\theta(\eta) = \nabla^2(\eta) &= (p\delta + \alpha)(p\delta\eta + \alpha\eta) \\
&= p\delta(p\delta\eta) + p\delta(\alpha\eta) + \alpha p\delta\eta + \alpha^2\eta \\
&= p\delta(p\delta\eta) + p\delta\alpha\eta + \alpha^2\eta \\
&= (p\delta p\delta p + p\delta\alpha + \alpha^2)(\eta) ,
\end{aligned} \tag{8.36}$$

since, by using $p\eta = p$ and $p^2 = p$, one has that

$$\begin{aligned}
p\delta(p\delta\eta) &= p\delta(p\delta(p\eta)) \\
&= p\delta(p\delta p\eta + p\delta\eta) \\
&= p\delta p\delta p\eta - p\delta p\delta\eta + p\delta p\delta\eta \\
&= p\delta p\delta p\eta .
\end{aligned} \tag{8.37}$$

With any connection ∇ on the module $\mathcal{E}$ there is associated a *dual connection* ∇' on the dual module $\mathcal{E}'$. Notice first, that there is a pairing

$$(\cdot, \cdot) : \mathcal{E}' \times \mathcal{E} \longrightarrow \mathcal{A} , \quad (\phi, \eta) =: \phi(\eta) , \tag{8.38}$$

which, due to (4.22), with respect to the right-module structure on $\mathcal{E}'$ (and taking $\mathcal{A}$ to be a *-algebra) has the following property

$$(\phi \cdot a, \eta \cdot b) = a^*(\phi, \eta)b , \quad \forall \phi \in \mathcal{E}', \eta \in \mathcal{E}, a, b \in \mathcal{A} . \tag{8.39}$$

Therefore, it can be extended to maps

$$\begin{aligned}
(\cdot, \cdot) : \mathcal{E}' \otimes_A \Omega\mathcal{A} \times \mathcal{E} &\longrightarrow \mathcal{A} , \quad (\phi \cdot \alpha, \eta) = \alpha^*(\phi, \eta) , \\
(\cdot, \cdot) : \mathcal{E}' \times \mathcal{E} \otimes_A \Omega\mathcal{A} &\longrightarrow \mathcal{A} , \quad (\phi, \eta \cdot \beta) = (\phi, \eta)\beta ,
\end{aligned} \tag{8.40}$$

for any $\phi \in \mathcal{E}'; \eta \in \mathcal{E}$; and $\alpha, \beta \in \Omega\mathcal{A}$.

Let us suppose now that we have a connection ∇ on $\mathcal{E}$. The dual connection

$$\nabla' : \mathcal{E}' \to \mathcal{E}' \otimes_A \Omega^1 \mathcal{A} , \tag{8.41}$$

is defined by

$$\delta(\phi, \eta) = -(\nabla'\phi, \eta) + (\phi, \nabla'\eta) , \quad \forall \phi \in \mathcal{E}', \eta \in \mathcal{E} . \tag{8.42}$$

It is easy to check the right-Leibniz rule

$$\nabla'(\phi \cdot a) = (\nabla'\phi)a + \phi \otimes_A \delta a , \quad \forall \, \phi \in \mathcal{E}' , a \in \mathcal{A} . \tag{8.43}$$

Indeed, for any $\phi \in \mathcal{E}', a \in \mathcal{A}$, and $\eta \in \mathcal{E}$, by using (8.39), (8.42), (7.29) and (8.40) respectively, we have

$$
\begin{aligned}
\delta(\phi \cdot a, \eta) &= -(\nabla'(\phi \cdot a), \eta) + (\phi \cdot a, \nabla'\eta) \\
\delta a^*(\phi, \eta) + a^* \delta(\phi, \eta) &= -(\nabla'(\phi \cdot a), \eta) + a^*(\phi, \nabla'\eta) \\
\delta a^*(\phi, \eta) - a^* \delta(\nabla'\phi, \eta) &= -(\nabla'(\phi \cdot a), \eta) \\
-(\delta a)^*(\phi, \eta) - a^* \delta(\nabla'\phi, \eta) &= -(\nabla'(\phi \cdot a), \eta) \\
(\phi \otimes_A \delta a, \eta) + ((\nabla'\phi) \cdot a, \eta) &= (\nabla'(\phi \cdot a), \eta) ,
\end{aligned}
\tag{8.44}
$$

from which (8.43) follows.

8.3 Connections Compatible with Hermitian Structures

Suppose now that we have a Hermitian structure $\langle \cdot, \cdot \rangle$ on the module $\mathcal{E}$ as defined in Sect. 4.3. A connection ∇ on $\mathcal{E}$ is said to be *compatible with the Hermitian structure* if the following condition is satisfied [32],

$$- \langle \nabla\eta, x \rangle + \langle \eta, \nabla\xi \rangle = \delta \langle \eta, \xi \rangle , \quad \forall \, \eta, \xi \in \mathcal{E} . \tag{8.45}$$

Here the Hermitian structure is extended to linear maps (denoted with the same symbol) : $\mathcal{E} \otimes_A \Omega^1 A \times \mathcal{E} \to \Omega^1 A$ and : $\Omega^1 A \otimes_A \mathcal{E} \times \mathcal{E} \to \Omega^1 A$ by

$$
\begin{aligned}
\langle \eta \otimes_A \omega, \xi \rangle &= \omega^* \langle \eta, \xi \rangle , \\
\langle \eta, \xi \otimes_A \omega \rangle &= \langle \eta, \xi \rangle \omega , \quad \forall \, \eta, \xi \in \mathcal{E} , \, \omega \in \Omega^1 A .
\end{aligned}
\tag{8.46}
$$

Also, the minus sign on the left hand side of eq. (8.45) is due to the choice $(\delta a)^* = -\delta a^*$ which we have made in (7.29).

Compatible connections always exist. As explained in Sect. 4.3, any Hermitian structure on $\mathcal{E} = p\mathcal{A}^N$ can be written as $\langle \eta, \xi \rangle = \sum_{j=1}^N \eta_j^* \xi_j$ with $\eta = p\eta = (\eta_1, \cdots, \eta_N)$ and the same for ξ. Then the Grassman connection (8.28) is compatible, since

$$
\begin{aligned}
\delta \langle \eta, \xi \rangle &= \delta(\sum_{j=1}^N \eta_j^* \xi_j) \\
&= \sum_{j=1}^N \delta\eta_j^* \, \xi_j + \sum_{j=1}^N \eta_j^* \, \delta\xi_j = -\sum_{j=1}^N (\delta\eta_j)^* \, \xi_j + \sum_{j=1}^N \eta_j^* \, \delta\xi_j \\
&= - \langle \delta\eta, p\xi \rangle + \langle p\eta, \delta\xi \rangle \\
&= - \langle p\delta\eta, \xi \rangle + \langle \eta, p\delta\xi \rangle \\
&= - \langle \nabla_0\eta, \xi \rangle + \langle \eta, \nabla_0\xi \rangle .
\end{aligned}
\tag{8.47}
$$

For a general connection (8.34), the compatibility with the Hermitian structure reduces to

$$\langle \alpha\eta, \xi \rangle - \langle \eta, \alpha\xi \rangle = 0 \ , \quad \forall \ \eta, \xi \in \mathcal{E} \ , \tag{8.48}$$

which just says that the gauge potential is Hermitian,

$$\alpha^* = \alpha \ . \tag{8.49}$$

We still use the symbol $CC(\mathcal{E})$ to denote the space of compatible universal connections on $\mathcal{E}$.

8.4 The Action of the Gauge Group

The group $\mathcal{U}(\mathcal{E})$ of unitary automorphisms of the module $\mathcal{E}$, defined in (4.29) plays the rôle of the *infinite dimensional group of gauge transformations*. Indeed, there is a natural action of such a group on the space $CC(\mathcal{E})$ of universal compatible connections on $\mathcal{E}$. It is given by

$$(u, \nabla) \longrightarrow \nabla^u =: u\nabla u^* \ , \quad \forall \ u \in \mathcal{U}(\mathcal{E}), \ \nabla \in CC(\mathcal{E}) \ . \tag{8.50}$$

It is then straightforward to check that the curvature transforms in a covariant way

$$(u, \theta) \longrightarrow \theta^u =: u\theta u^* \ , \tag{8.51}$$

since, evidently, $\theta^u = (\nabla^u)^2 = u\nabla u^* u\nabla u^* = u\nabla^2 u^* = u\theta u^*$.
As for the gauge potential, one has the usual affine transformation

$$(u, \alpha) \longrightarrow \alpha^u =: up\delta u^* + u\alpha u^* \ . \tag{8.52}$$

Indeed, for any $\eta \in \mathcal{E}$,

$$
\begin{aligned}
\nabla^u(\eta) &= u(p\delta + \alpha)u^*\eta = up\delta(u^*\eta) + u\alpha u^*\eta \\
&= pu(u^*\delta\eta) + up(\delta u^* \ \eta) + u\alpha u^* \quad \text{using} \quad up = pu \\
&= p\delta\eta + (up\delta u^* + u\alpha u^*)\eta \\
&= (p\delta + \alpha^u)\eta \ ,
\end{aligned}
\tag{8.53}
$$

which yields (8.52) for the transformed potential.

8.5 Connections on Bimodules

In constructing gravity theories one needs to introduce the analogues of linear connections. These are connections defined on the bimodule of 1-forms which plays the rôle of the cotangent bundle. Since this module is in fact a bimodule, it seems natural to exploit both left and right module structures. In fact, to discuss reality conditions and curvature invariants the bimodule structure is crucial.

We refer to [54] for a theory of connection on central bimodules. An alternative idea which has been proposed (in [107] for linear connections and in [52] for the general case) is that of a 'braiding' which, by generalizing the permutation of forms, flips two elements of a tensor product so as to make possible a *left* Leibniz rule once a *right* Leibniz rule is satisfied.

Let $\mathcal{E}$ be an $\mathcal{A}$-bimodule which is left and right projective, and endowed with a right connection, i.e. a linear map $\nabla : \mathcal{E} \to \mathcal{E} \otimes_A \Omega^1 A$ which obeys the right Leibniz rule (8.16).

Definition 8.5.1. *Given a bimodule isomorphism,*

$$\sigma : \Omega^1 A \otimes_A \mathcal{E} \longrightarrow \mathcal{E} \otimes_A \Omega^1 A , \tag{8.54}$$

the couple (∇, σ) is said to be compatible *if and only if a left Leibniz rule of the form*

$$\nabla(a\eta) = a(\nabla\eta) + \sigma(\delta a \otimes_A \eta) , \quad \forall\, a \in \mathcal{A} , \ \eta \in \mathcal{E} . \tag{8.55}$$

is satisfied.

We see that the rôle of the map σ is to bring the one form δa to the 'right place'. Notice that in general σ need not square to the identity, $\sigma \circ \sigma \neq \mathbb{I}$. In [52] σ is identified as the symbol of the connection.

To get a bigger space of connections a weaker condition has been proposed in [41] where the compatibility condition has been required to be satisfied only on the center of the bimodule. We recall, first of all, that the center $\mathcal{Z}(\mathcal{E})$ of a bimodule $\mathcal{E}$ is the bimodule defined by

$$\mathcal{Z}(\mathcal{E}) =: \{\eta \in \mathcal{E} \mid a\eta = \eta a , \ \forall a \in \mathcal{A}\} . \tag{8.56}$$

Now, let ∇^L be a *left connection*, namely a linear map : $\mathcal{E} \to \Omega^1 A \otimes_A \mathcal{E}$ satisfying the left Leibniz rule

$$\nabla^L(a\eta) = a\nabla^L\eta + \delta a \otimes_A \eta , \quad \forall\, a \in A, \ \eta \in E . \tag{8.57}$$

Likewise let ∇^R be a right connection, i.e. a linear map : $\mathcal{E} \to \mathcal{E} \otimes_A \Omega^1 A$ satisfying the right Leibniz rule

$$\nabla^R(\eta a) = (\nabla^R\eta)a + \eta \otimes_A \delta a , \quad \forall\, a \in A, \ \eta \in E. \tag{8.58}$$

Definition 8.5.2. *With σ a bimodule isomorphism as in (8.54), a pair (∇^L, ∇^R) is said to be σ-compatible if and only if*

$$\nabla^R \eta = (\sigma \circ \nabla^L)\eta , \quad \forall \, \eta \in \mathcal{Z}(\mathcal{E}) . \tag{8.59}$$

By requiring that the condition $\nabla^R = \sigma \circ \nabla^L$ be satisfied on the whole bimodule $\mathcal{E}$, one can equivalently think of a pair (∇^L, ∇^R) as a right connection ∇^R fulfilling the additional left Leibniz rule (8.55) and so reproducing the previously described situation.[1]

It should be mentioned that whereas Definition 8.5.1 is compatible with tensor products over the algebra, Definition 8.5.2 does not enjoy this property.

Several examples of gravity and Kaluza-Klein theories which use the bimodule structure of $\mathcal{E} = \Omega^1 \mathcal{A}$ have been constructed: see [100, 103] and references therein.

In Chap. 10, we shall describe gravity theories which use only one structure (the right one, although it would be completely equivalent to use the left one). In this context, the usual Einstein gravity has been obtained as a particular case.

---

[1] In [40] a connection on a bimodule is also defined as a pair consisting of a left and right connection. There, however, there is no σ-compatibility condition while the additional conditions of ∇^L being a right $\mathcal{A}$-homomorphism and ∇^R being a left $\mathcal{A}$-homomorphism are imposed. These latter conditions, are not satisfied in the classical commutative case where $\mathcal{Z}(\mathcal{E}) = \mathcal{E} = \Omega^1(M)$.

9. Field Theories on Modules

In this section we shall describe how to construct field theoretical models in the algebraic noncommutative framework developed by Connes. Throughout this Section, the basic ingredient will be a spectral triple $(\mathcal{A}, \mathcal{H}, D)$ of dimension n. Associated with it there is the algebra $\Omega_D\mathcal{A} = \oplus_p \Omega_D^p\mathcal{A}$ of forms, with exterior differential d, as constructed in Sect. 7.2.

9.1 Yang-Mills Models

The theory of connections on any (finite projective) $\mathcal{A}$-module $\mathcal{E}$, with respect to the differential calculus $(\Omega_D\mathcal{A}, d)$ is, *mutatis mutandis*, formally the same as the theory of universal connections developed in Sect. 8.2.

Definition 9.1.1. *A connection on the $\mathcal{A}$-module $\mathcal{E}$ is a $\mathbb{C}$-linear map*

$$\nabla : \mathcal{E} \otimes_\mathcal{A} \Omega_D^p\mathcal{A} \longrightarrow \mathcal{E} \otimes_\mathcal{A} \Omega_D^{p+1}\mathcal{A} , \tag{9.1}$$

satisfying the Leibniz rule

$$\nabla(\omega\rho) = (\nabla\omega)\rho + (-1)^p\omega d\rho , \quad \forall \, \omega \in \mathcal{E} \otimes_\mathcal{A} \Omega_D^p\mathcal{A} , \, \rho \in \Omega_D\mathcal{A} . \tag{9.2}$$

The composition $\nabla^2 = \nabla \circ \nabla : \mathcal{E} \otimes_\mathcal{A} \Omega_D^p\mathcal{A} \to \mathcal{E} \otimes_\mathcal{A} \Omega_D^{p+2}\mathcal{A}$ is $\Omega_D\mathcal{A}$-linear and its restriction to $\mathcal{E}$ is the *curvature* $F : \mathcal{E} \to \mathcal{E} \otimes_\mathcal{A} \Omega_D^2\mathcal{A}$ of the connection. The curvature is $\mathcal{A}$ linear, $F(\eta a) = F(\eta)a$, for any $\eta \in \mathcal{E}$, and $a \in \mathcal{A}$, and satisfies,

$$\nabla^2(\eta \otimes_\mathcal{A} \rho) = F(\eta)\rho , \quad \forall \, \eta \in \mathcal{E} , \, \rho \in \Omega_D\mathcal{A} . \tag{9.3}$$

As before, thinking of the curvature F as an element of $End_\mathcal{A}\mathcal{E} \otimes_\mathcal{A} \Omega_D^2\mathcal{A}$, it satisfies the Bianchi identity,

$$[\nabla, F] = 0 . \tag{9.4}$$

As was mentioned before, connections always exist on a projective module. If $\mathcal{E} = p\mathcal{A}^N$, it is possible to write any connection as

$$\nabla = pd + A , \tag{9.5}$$

where A is any element in $End_A \mathcal{E} \otimes_A \Omega_D^1 A \simeq \mathbb{M}_A(A) \otimes_A \Omega_D^1 A$ such that $A = Ap = pA = pAp$. The matrix of 1-forms A is called the *gauge potential* of the connection ∇. For the corresponding curvature F we have

$$F = pdA + A^2 + pdpdp . \tag{9.6}$$

The space $C(\mathcal{E})$ of all connections on $\mathcal{E}$ is an affine space modelled on the space $End_A \mathcal{E} \otimes_A \Omega_D^1 A$.

The compatibility of the connection ∇ with respect to a Hermitian structure on $\mathcal{E}$ is expressed exactly as in Sect. 8.3,

$$- \langle \nabla \eta, \xi \rangle + \langle \eta, \nabla \xi \rangle = d \langle \eta, \xi \rangle , \quad \forall \, \eta, \xi \in \mathcal{E} . \tag{9.7}$$

As before, the Hermitian structure is extended to linear maps

$$\begin{aligned}
\mathcal{E} \otimes_A \Omega_D^1 A \times \mathcal{E} \to \Omega_D^1 A , & \quad \langle \eta \otimes_A \omega, \xi \rangle = \omega^* \langle \eta, \xi \rangle , \\
\Omega_D^1 A \otimes_A \mathcal{E} \times \mathcal{E} \to \Omega_D^1 A , & \quad \langle \eta, \xi \otimes_A \omega \rangle = \langle \eta, x \rangle \omega , \\
& \quad \forall \, \eta, \xi \in \mathcal{E}, \, \omega \in \Omega_D^1 A . \tag{9.8}
\end{aligned}$$

The connection (9.5) is compatible with the Hermitian structure $\langle \eta, \xi \rangle = \sum_{j=1}^N \eta_j^* \xi_j$ on $\mathcal{E} = pA^N$ ($\eta = (\eta_1, \cdots, \eta_N) = p\eta$ and the same for ξ), provided the gauge potential is Hermitian,

$$A^* = A . \tag{9.9}$$

The action of the group $\mathcal{U}(\mathcal{E})$ of unitary automorphisms of the module $\mathcal{E}$ on the space $C(\mathcal{E})$ of compatible connections on $\mathcal{E}$ is given by

$$(u, \nabla) \longrightarrow \nabla^u =: u \nabla u^* , \quad \forall \, u \in \mathcal{U}(\mathcal{E}), \, \nabla \in C(\mathcal{E}) . \tag{9.10}$$

Hence, the gauge potential and the curvature transform in the usual way

$$(u, A) \longrightarrow A^u = u[pd + A]u^* , \tag{9.11}$$

$$(u, F) \longrightarrow F^u =: u F u^* , \quad \forall \, u \in \mathcal{U}(\mathcal{E}). \tag{9.12}$$

The following proposition clarifies in which sense the connections defined in 8.2.1 are universal.

Proposition 9.1.1. *The representation π in (7.37) can be extended to a surjective map*

$$\mathbb{I} \otimes \pi : CC(\mathcal{E}) \longrightarrow C(\mathcal{E}) ; \tag{9.13}$$

consequently, any compatible connection is the composition of π with a universal compatible connection.

Proof. By construction, π is a surjection from $\Omega^1 A$ to $\pi(\Omega^1 A) \simeq \Omega_D^1 A$. Then, we get a surjection $\mathbb{I} \otimes \pi : End_A \mathcal{E} \otimes_A \Omega^1 A \to End_A \mathcal{E} \otimes_A \Omega_D^1 A$. Finally, define $\mathbb{I} \otimes \pi(p \circ \delta) = p \circ d$ to get the desired surjection $: CC(\mathcal{E}) \longrightarrow C(\mathcal{E})$.

By using the Hermitian structure on $\mathcal{E}$ together with an ordinary matrix trace over 'internal indices', one can construct an inner product on $End_A\mathcal{E}$. By combining this product with the inner product on $\Omega_D^2\mathcal{A}$ given in (7.92), one then obtains a natural inner product $\langle\ ,\ \rangle_2$ on the space $End_A\mathcal{E}\otimes_A\Omega_D^2\mathcal{A}$. Since the curvature F is an element of such a space, the following definition makes sense.

Definition 9.1.2. *The Yang-Mills action for the connection* ∇ *with curvature* F *is given by*

$$YM(\nabla) = \langle F, F\rangle_2 \ . \tag{9.14}$$

By its very construction it is invariant under the gauge transformations of (9.11) and (9.12).

Consider now the tensor product $\mathcal{E}\otimes_A\mathcal{H}$. This space can be promoted to a Hilbert space by combining the Hermitian structure on $\mathcal{E}$ with the scalar product on $\mathcal{H}$,

$$(\eta_1\otimes_A\psi_1, \eta_2\otimes_A\psi_2) =: (\psi_1, \langle\eta_1, \eta_2\rangle\,\psi_2)\ ,\quad \forall\ \eta_1, \eta_2 \in \mathcal{E},\ \ \psi_1,\psi_2 \in \mathcal{H}\ . \tag{9.15}$$

By using the projection (7.37) we get a projection

$$\mathbb{I}_\mathcal{E}\otimes\pi : \mathcal{E}\otimes_A\Omega_D\mathcal{A} \longrightarrow \mathcal{B}(\mathcal{E}\otimes_A\mathcal{H})\ , \tag{9.16}$$

and an inner product on $(\mathbb{I}_\mathcal{E}\otimes\pi)(\mathcal{E}\otimes_A\Omega^p\mathcal{A})$ given by

$$\langle T_1, T_2\rangle_p = tr_\omega T_1^* T_2 |\mathbb{I}_\mathcal{E}\otimes D|^{-n}\ , \tag{9.17}$$

which is the analogue of the inner product (7.92). The corresponding orthogonal projector P has a range which can be identified with $\mathcal{E}\otimes_A\Omega_D^p\mathcal{A}$.

If $\nabla_{un} \in CC(\mathcal{E})$ is any universal connection with curvature θ_{un}, one defines a pre-Yang-Mills action $I(\nabla_{un})$ by,

$$I(\nabla_{un}) = tr_\omega\pi(\theta_{un})^2 |\mathbb{I}\otimes D|^{-n}\ . \tag{9.18}$$

One has the analogue of proposition (8.1.1).

Proposition 9.1.2. *For any compatible connection* $\nabla \in C(\mathcal{E})$, *one has that*

$$YM(\nabla) = inf\{I(\nabla_{un})\mid \pi(\nabla_{un}) = \nabla\}\ . \tag{9.19}$$

Proof. The proof is analogous to that of Proposition 8.1.1.

It is also possible to define a *topological action* and extend the usual inequality between Chern classes of vector bundles and the value of the Yang-Mill action on an arbitrary connection on the bundle. First observe that, from definition (9.14) of the Yang-Mills action functional, if D is replaced by λD, then $YM(\nabla)$ is replaced by $|\lambda|^{4-n}YM(\nabla)$. Therefore, it has a chance of being related to 'topological invariants' of finite projective modules only if $n = 4$. Let us then assume that our spectral triple is four dimensional. We also need it to be even with a $\mathbb{Z}_2$ grading Γ. With these ingredients, one defines two traces on the algebra $\Omega^4 \mathcal{A}$,

$$\tau(a_0 \delta a_1 \cdots \delta a_4) = tr_w(a_0[D, a_1] \cdots [D, a_1]|D|^{-4}) , \quad a_j \in \mathcal{A}$$
$$\Phi(a_0 \delta a_1 \cdots \delta a_4) = tr_w(\Gamma a_0[D, a_1] \cdots [D, a_1]|D|^{-4}) , \quad a_j \in \mathcal{A} . \quad (9.20)$$

By using the projection (9.16) and an 'ordinary trace over internal indices' and by substituting Γ with $\mathbb{I}_{\mathcal{E}} \otimes \Gamma$ and $|D|^{-4}$ with $\mathbb{I}_{\mathcal{E}} \otimes |D|^{-4}$, the previous traces can be extended to traces $\widetilde{\tau}$ and $\widetilde{\Phi}$ on $End_{\mathcal{A}}\mathcal{E} \otimes_{\mathcal{A}} \Omega^4 \mathcal{A}$. Then, by the definition (9.18), one has that

$$I(\nabla_{un}) = \widetilde{\tau}(\theta_{un}^2) , \quad \forall \nabla_{un} \in CC(\mathcal{E}) , \quad (9.21)$$

with θ_{un} the curvature of ∇_{un}. Furthermore, since the operator $\mathbb{I}_{\mathcal{E}} \pm \Gamma$ is positive and anticommutes with $\pi(\Omega^4 \mathcal{A})$,[1] one can establish an inequality [32]

$$\widetilde{\tau}(\theta_{un}^2) \geq |\widetilde{\Phi}(\theta_{un}^2)| , \quad \forall \nabla_{un} \in CC(\mathcal{E}) . \quad (9.22)$$

In turn, by using (9.21) and (9.1.2), one gets the inequality

$$YM(\nabla) \geq |\widetilde{\Phi}(\theta_{un}^2)| , \quad \pi(\nabla_{un}) = \nabla . \quad (9.23)$$

It turns out that $\widetilde{\Phi}(\theta_{un}^2)$ is a closed *cyclic cocycle* and its topological interpretation in terms of topological invariants of finite projective modules follows from the pairing between K-theory and cyclic cohomology. Indeed, the value of $\widetilde{\Phi}$ does not depend on the particular connection and one could evaluate it on the curvature $\theta_0 = pdpdp$ of the Grassmannian connection. Moreover, it depends only on the stable isomorphism class $[p] \in K_0(\mathcal{A})$. We refer to [32] for details. In the next section, we shall show that for the canonical triple over an ordinary four dimensional manifold, the term $\widetilde{\Phi}(\theta_{un}^2)$ reduces to the usual topological action.

9.1.1 Usual Gauge Theory

For simplicity we shall consider the case when $n = 4$. For the canonical triple $(\mathcal{A}, \mathcal{H}, D, \Gamma)$ over the (four dimensional) manifold M, as described in Sect. 6.5, consider a matrix A of usual 1-forms and a universal connection $\nabla = p\delta + \alpha$ such that $\sigma_1(\pi(\alpha)) = \gamma(A)$. Then $P(\pi(\theta)) = P(\pi(\delta\alpha + \alpha^2)) = \gamma(F)$ with $F = d_M A + A \wedge A$. On making use of eq. (7.96), with an additional matrix trace over the 'internal indices', we get

$$YM(A) = inf\{I(\alpha) \mid \pi(\alpha) = A\}$$
$$= \frac{1}{8\pi^2} \int_M ||F||^2 dx . \quad (9.24)$$

---

[1] We recall that Γ commutes with elements of $\mathcal{A}$ and anticommutes with D.

This is the usual Yang-Mills action for the gauge potential A.
More explicitly, let $\alpha = \sum_j f_j \delta g_j$. Then, we have

$$\pi(\alpha) = \gamma^\mu A_\mu , \quad A_\mu = \sum_j f_j \partial_\mu g_j ,$$

$$P(\pi(\delta\alpha + \alpha^2)) = \gamma^{\mu\nu} F_{\mu\nu} , \quad \gamma^{\mu\nu} = \frac{1}{2}(\gamma^\mu \gamma^\nu - \gamma^\nu \gamma^\mu) . \quad (9.25)$$

By using the trace theorem 6.3.1 again, (with an additional matrix trace Tr over the 'internal indices') one gets

$$
\begin{aligned}
YM(A) &=: \frac{1}{8\pi^2} \int_M tr(\gamma^{\mu\nu}\gamma^{\rho\sigma}) Tr(F_{\mu\nu} F_{\rho\sigma}) dx \\
&=: \frac{1}{8\pi^2} \int_M g^{\mu\sigma} g^{\nu\rho} Tr(F_{\mu\nu} F_{\rho\sigma}) dx \\
&=: \frac{1}{8\pi^2} \int_M Tr(F \wedge *F) .
\end{aligned}
\quad (9.26)
$$

With the same token, we get for the topological action

$$
\begin{aligned}
Top(A) &=: \frac{1}{8\pi^2} \int_M tr(\Gamma\gamma^{\mu\nu}\gamma^{\rho\sigma}) Tr(F_{\mu\nu} F_{\rho\sigma}) dx \\
&=: -\frac{1}{8\pi^2} \int_M \varepsilon^{\mu\nu\rho\sigma} Tr(F_{\mu\nu} F_{\rho\sigma}) dx \\
&=: -\frac{1}{8\pi^2} \int_M Tr(F \wedge F) ,
\end{aligned}
\quad (9.27)
$$

which is the usual topological action.
Here we have used the following (normalized) traces of gamma matrices

$$tr(\gamma^\mu\gamma^\nu\gamma^\rho\gamma^\sigma) = (g^{\mu\nu} g^{\rho\sigma} - g^{\mu\rho} g^{\nu\sigma} + g^{\mu\sigma} g^{\nu\rho}) \quad (9.28)$$
$$tr(\Gamma\gamma^\mu\gamma^\nu\gamma^\rho\gamma^\sigma) = -\varepsilon^{\mu\nu\rho\sigma} . \quad (9.29)$$

9.1.2 Yang-Mills on a Two-Point Space

We shall first study all modules on the two-point space $Y = \{1,2\}$ described in Sect. 6.8. The associated algebra is $\mathcal{A} = \mathbb{C} \oplus \mathbb{C}$. The generic module $\mathcal{E}$ will be of the form $\mathcal{E} = p\mathcal{A}^{n_1}$, with n_1 a positive integer, and p a $n_1 \times n_1$ idempotent matrix with entries in $\mathcal{A}$. The most general such idempotent can be written as a diagonal matrix of the form

$$p = \mathrm{diag}[\underbrace{(1,1), \cdots, (1,1)}_{n_1}, \underbrace{(1,0), \cdots, (1,0),}_{n_1-n_2}] , \quad (9.30)$$

with $n_2 \leq n_1$. Therefore, the module $\mathcal{E}$ can be thought of as n_1 copies of $\mathbb{C}$ on the point 1 and n_2 copies of $\mathbb{C}$ on the point 2,

$$\mathcal{E} = \mathbb{C}^{n_1} \oplus \mathbb{C}^{n_2} \ . \tag{9.31}$$

The module is trivial if and only if $n_1 = n_2$. There is a *topological number* which measures the triviality of the module and that, in this case, turns out to be proportional to $n_1 - n_2$. From eq. (9.6), the curvature of the Grassmannian connection on $\mathcal{E}$ is just $F_0 = pdpdp$. The aforementioned topological number is then

$$c(\mathcal{E}) =: tr\Gamma F_0^2 = tr\Gamma(pdpdp)^2 = tr\Gamma p(dp)^4 \ . \tag{9.32}$$

Here Γ is the grading matrix given by (6.81) and, as the spectral triple is 0-dimensional, the Dixmier trace reduces to an ordinary trace.[2] This is really the same as the topological action $\Phi(\theta_{un}^2)$ encountered in Sect. 9.1. It takes a little algebra to find that, for a module of the form (9.31), one has

$$c(\mathcal{E}) = tr(M^*M)^4(n_1 - n_2) \ , \tag{9.33}$$

where M is the matrix appearing in the corresponding operator D as in (6.80).

Let us now turn to gauge theories. First recall that from the analysis of Sect. 7.2.2 there are no junk forms and that Connes' forms are the images of universal forms through π, $\Omega_D\mathcal{A} = \pi(\Omega\mathcal{A})$ with π injective. We shall consider the simple case of a 'trivial 1-dimensional bundle' over Y, namely we shall take as the module of sections just $\mathcal{E} = \mathcal{A}$. A vector potential is then a self-adjoint element $A \in \Omega_D^1\mathcal{A}$ and is determined by a complex number $\Phi \in \mathbb{C}$,

$$A = \begin{bmatrix} 0 & \overline{\Phi}M^* \\ \Phi M & 0 \end{bmatrix} \ . \tag{9.34}$$

If α is the universal form such that $\pi(\alpha) = A$, then

$$\alpha = -\overline{\Phi}e\delta e - \Phi(1-e)\delta(1-e) \ , \tag{9.35}$$

and its curvature is

$$\delta\alpha + \alpha^2 = -(\overline{\Phi} + \Phi + |\Phi|^2)\delta e\delta e \ . \tag{9.36}$$

Finally, the Yang-Mills curvature turns out to be

$$YM(A) =: tr\pi(\delta\alpha + \alpha^2)^2 = 2tr(M^*M)^2 \ (|\Phi + 1|^2 - 1)^2 \ . \tag{9.37}$$

The gauge group $\mathcal{U}(\mathcal{E})$ is the group of unitary elements of $\mathcal{A}$, namely the group $\mathcal{U}(\mathcal{E}) = U(1) \times U(1)$. Its elements can be represented as diagonal matrices. Indeed, for $u \in U(1) \times U(1)$,

$$u = \begin{bmatrix} u_1 & 0 \\ 0 & u_2 \end{bmatrix} \ , \quad |u_1|^2 = 1 \ , \quad |u_2|^2 = 1 \ . \tag{9.38}$$

---

[2] In fact, in (9.32), Γ is really $\mathbb{I} \otimes \Gamma$.

Its action, $A^u = uAu^* + udu^*$, on the gauge potential results in multiplication by $u_1^* u_2$ on the variable $\Phi + 1$,

$$(\Phi + 1)^u = (\Phi + 1)u_1^* u_2 , \tag{9.39}$$

and the action (9.37) is gauge invariant.

We see that in this example the action, $YM(A)$, reproduces the usual situation of broken symmetry for the 'Higgs field' $\Phi + 1$: there is a S^1-worth of minima which are acted upon nontrivially by the gauge group. This fact has been used in [38] in a reconstruction of the Standard Model. The Higgs field has a geometrical interpretation: it is the component of a gauge connection along an 'internal' discrete direction made up of two points.[3]

9.2 The Bosonic Part of the Standard Model

There are excellent review papers on the derivation of the Standard Model using noncommutative geometry. In particular one should consult [130, 102] and [86] and we do not feel the need to add more to these. Rather we shall only give an overview of the main features. Here we limit ourselves to the bosonic content of the model while postponing to the following sections the description of the fermionic part.

In [38], Connes and Lott computed the Yang-Mills action $YM(\nabla)$ for a space which is the product of a Riemannian spin manifold M by a 'discrete' internal space Y consisting of two points. One constructs the product, as described in Sect. 6.9, of the canonical triple $(C^\infty(M), L^2(M, S), D_S, \Gamma_5)$ on a Riemannian four dimensional spin manifold with the finite zero dimensional triple $(\mathbb{C} \oplus \mathbb{C}, \mathcal{H}_1 \oplus \mathcal{H}_2, D_F)$ described in Sects. 6.8 and 9.1.2. The product triple is then given by

$$\mathcal{A} =: C^\infty(M) \otimes (\mathbb{C} \oplus \mathbb{C}) \simeq C^\infty(M) \oplus C^\infty(M) ,$$
$$\mathcal{H} =: L^2(M, S) \otimes (\mathcal{H}_1 \oplus \mathcal{H}_2) \simeq L^2(M, S) \otimes \mathcal{H}_1 \oplus L^2(M, S) \otimes \mathcal{H}_2 ,$$
$$D =: D_S \otimes \mathbb{I} + \Gamma_5 \otimes D_F . \tag{9.40}$$

A nice feature of the model is that one has a geometric interpretation of the Higgs field which appears as the component of the gauge field in the internal direction. Geometrically one has a space $M \times Y$ with two sheets which are at a distance of the order of the inverse of the mass scale of the theory (which appears in the operator D_F for the finite part as the parameter M). The differentiation in the space $M \times Y$ consists of differentiation on each copy of M together with a finite difference operation in the Y direction. A gauge

---

[3] One should mention that the first noncommutative extension of classical gauge theories in which the Higgs fields appear as components of the gauge connection in the 'noncommutative directions' was produced in the framework of the derivation based calculus [50].

potential A decomposes as a sum of an ordinary differential part $A^{(1,0)}$ and a finite difference part $A^{(0,1)}$ which turns out to be the Higgs field.

To get the full bosonic standard model one has to take for the finite part the algebra [34]

$$\mathcal{A}_F = \mathbb{C} \oplus \mathbb{H} \oplus \mathbb{M}_3(\mathbb{C}) \ , \tag{9.41}$$

where $\mathbb{H}$ is the algebra of quaternions. The unitary elements of this algebra form the group $U(1) \times SU(2) \times U(3)$. The finite Hilbert space $\mathcal{H}_F$ is the fermion space of leptons, quarks and their antiparticles $\mathcal{H}_F = \mathcal{H}_F^+ \oplus \mathcal{H}_F^- = \mathcal{H}_\ell^+ \oplus \mathcal{H}_q^+ \oplus \mathcal{H}_{\bar{\ell}}^- \oplus \mathcal{H}_{\bar{q}}^-$. As for the finite Dirac operator D_F is given by

$$D_F = \begin{bmatrix} Y & 0 \\ 0 & \overline{Y} \end{bmatrix} \ , \tag{9.42}$$

with Y an off-diagonal matrix which contains the Yukawa couplings. The real structure J_F defined by

$$J_F \left(\frac{\xi}{\eta} \right) = \left(\frac{\eta}{\xi} \right) \ , \quad \forall \ (\xi, \eta) \in \mathcal{H}_F^+ \oplus \mathcal{H}_F^- \ , \tag{9.43}$$

exchanges fermions with antifermions and it is such that

$$\begin{aligned} J_F^2 &= \mathbb{I} \ , \\ \Gamma_F J_F + J_F \Gamma_F &= 0 \ , \\ D_F J_F - J_F D_F &= 0 \ . \end{aligned} \tag{9.44}$$

Next, one defines an action of the algebra (9.41) so as to meet the other requirements in the Definition 6.7.1 of a real structure. For the details on this we refer to [34, 102] as well as for the details on the construction of the full bosonic Standard Model action starting from the Yang-Mills action $YM(\nabla)$ on a 'rank one trivial' module associated with the product geometry

$$\begin{aligned} \mathcal{A} &=: C^\infty(M) \otimes \mathcal{A}_F \ , \\ \mathcal{H} &=: L^2(M, S) \otimes \mathcal{H}_F \ , \\ D &=: D_S \otimes \mathbb{I} + \Gamma_5 \otimes D_F \ . \end{aligned} \tag{9.45}$$

The product triple has a real structure given by

$$J = C \otimes J_F \ , \tag{9.46}$$

with C the charge-conjugation operation on $L^2(M, S)$ and J_F the real structure of the finite geometry.

The final model has problems, notably unrealistic mass relations [102] and a disturbing fermion doubling [98]. It is worth mentioning that while the standard model can be obtained from noncommutative geometry, most models of the Yang-Mills-Higgs type cannot [125, 77, 97].

9.3 The Bosonic Spectral Action

Recently, in [36], Connes has proposed a new interpretation of gauge degrees of freedom as the 'inner fluctuations' of a noncommutative geometry. These fluctuations replace the operator D, which gives the 'external geometry', by $D + A + JAJ^*$, where A is the gauge potential and J is the real structure. In fact, there is also a purely geometrical (spectral) action, depending only on the spectrum of the operator D, which, for a suitable algebra (noncommutative geometry of the Standard Model) gives the Standard Model Lagrangian coupled to gravity.

Firstly, we recall that if M is a smooth (paracompact) manifold, then the group $Diff(M)$ of diffeomorphisms of M, is isomorphic with the group $Aut(C^\infty(M))$ of (*-preserving) automorphisms of the algebra $C^\infty(M)$ [1]. Here $Aut(C^\infty(M))$ is the collection of all invertible, linear maps α from $C^\infty(M)$ into itself such that $\alpha(fg) = \alpha(f)\alpha(g)$ and $\alpha(f^*) = (\alpha(f))^*$, for any $f, g \in C^\infty(M)$; $Aut(C^\infty(M))$ is a group under map composition. The relation between a diffeomorphism $\varphi \in Diff(M)$ and the corresponding automorphism $\alpha_\varphi \in Aut(C^\infty(M))$ is via pull-back

$$\alpha_\varphi(f)(x) =: f(\varphi^{-1}(x)) , \quad \forall f \in C^\infty(M) , \ x \in M . \tag{9.47}$$

If $\mathcal{A}$ is any noncommutative algebra (with unit) one defines the group $Aut(\mathcal{A})$ exactly as before and $\varphi(\mathbb{I}) = \mathbb{I}$, for any $\varphi \in Aut(\mathcal{A})$. This group is the analogue of the group of diffeomorphism of the (virtual) noncommutative space associated with $\mathcal{A}$. Now, with any element u of the unitary group $\mathcal{U}(\mathcal{A})$ of $\mathcal{A}$, $\mathcal{U}(\mathcal{A}) = \{u \in \mathcal{A} , \ uu^* = u^*u = \mathbb{I}\}$, there is an *inner automorphism* $\alpha_u \in Aut(\mathcal{A})$ defined by

$$\alpha_u(a) = uau^* , \quad \forall a \in \mathcal{A} . \tag{9.48}$$

One can easily convince oneself that $\alpha_{u^*} \circ \alpha_u = \alpha_u \circ \alpha_{u^*} = \mathbb{I}_{Aut(\mathcal{A})}$, for any $u \in \mathcal{U}(\mathcal{A})$. The subgroup $Inn(\mathcal{A}) \subset Aut(\mathcal{A})$ of all inner automorphisms is a normal subgroup. First of all, any automorphism will preserve the group of unitaries in $\mathcal{A}$. If $u \in \mathcal{U}(\mathcal{A})$ and $\varphi \in Aut(\mathcal{A})$, then $\varphi(u)(\varphi(u))^* = \varphi(u)\varphi(u^*) = \varphi(uu^*) = \varphi(\mathbb{I}) = \mathbb{I}$; analogously $(\varphi(u))^*\varphi(u) = \mathbb{I}$ and $\varphi(u) \in \mathcal{U}(\mathcal{A})$. Furthermore,

$$\alpha_{\varphi(u)} = \varphi \circ \alpha_u \circ \varphi^{-1} \in Inn(\mathcal{A}) , \quad \forall \varphi \in Aut(\mathcal{A}) , \ \alpha_u \in Inn(\mathcal{A}) . \tag{9.49}$$

Indeed, with $a \in \mathcal{A}$, for any $\varphi \in Aut(\mathcal{A})$ and $\alpha_u \in Inn(\mathcal{A})$ one finds

$$
\begin{aligned}
\alpha_{\varphi(u)}(a) &= \varphi(u)a\varphi(u^*) \\
&= \varphi(u)\varphi(\varphi^{-1}(a)\varphi(u^*) \\
&= \varphi(u\varphi^{-1}(a)u^*) \\
&= (\varphi \circ \alpha_u \circ \varphi^{-1})(a) ,
\end{aligned}
\tag{9.50}
$$

from which one gets (9.49).

We denote by $Out(\mathcal{A}) =: Aut(\mathcal{A})/Inn(\mathcal{A})$ the outer automorphisms. There is a short exact sequence of groups

$$\mathbb{I}_{Aut(\mathcal{A})} \longrightarrow Inn(\mathcal{A}) \longrightarrow Aut(\mathcal{A}) \longrightarrow Out(\mathcal{A}) \longrightarrow \mathbb{I}_{Aut(\mathcal{A})} . \quad (9.51)$$

For any commutative $\mathcal{A}$ (in particular for $\mathcal{A} = C^{\infty}(M)$) there are no non-trivial inner automorphisms and $Aut(\mathcal{A}) \equiv Out(\mathcal{A})$ (in particular $Aut(\mathcal{A}) \equiv Out(\mathcal{A}) \simeq Diff(M)$).

The interpretation that emerges is that the group $Inn(\mathcal{A})$ will give 'internal' gauge transformations while the group $Out(\mathcal{A})$ will provide 'external' diffeomorphisms. In fact, gauge degrees of freedom are the 'inner fluctuations' of the noncommutative geometry. This is due to the following beautiful fact. Consider the real triple $(\mathcal{A}, \mathcal{H}, \pi, D)$, where we have explicitly indicated the representation π of the algebra $\mathcal{A}$ on the Hilbert space $\mathcal{H}$. The real structure is provided by the antilinear isometry J with properties as in Definition 6.7.1. Any inner automorphism $\alpha_u \in Inn(\mathcal{A})$ will produce a new representation $\pi_u =: \pi \circ \alpha_u$ of $\mathcal{A}$ in $\mathcal{H}$. It turns out that the replacement of the representation is equivalent to the replacement of the operator D by

$$D_u = D + A + \varepsilon' JAJ^* , \quad (9.52)$$

where $A = u[D, u^*]$ and $\varepsilon' = \pm 1$ from (6.75) according to the dimension of the triple. If the dimension is four, then $\varepsilon' = 1$; in what follows we shall limit ourselves to this case, the generalization being straightforward.

This result is so important and beautiful that we shall restate it as a Proposition.

Proposition 9.3.1. *For any inner automorphism $\alpha_u \in Inn(\mathcal{A})$, with u unitary, the triples $(\mathcal{A}, \mathcal{H}, \pi, D, J)$ and $(\mathcal{A}, \mathcal{H}, \pi \circ \alpha_u, D + u[D, u^*] + Ju[D, u^*]J^*, J)$ are equivalent, the intertwining unitary operator being given by*

$$U = uJuJ^* . \quad (9.53)$$

Proof. Note first that

$$UJU^* = J . \quad (9.54)$$

Indeed, by using the properties, from the Definition 6.7.1, of a real structure, we have,

$$
\begin{aligned}
UJU^* &= uJuJ^* JJu^* J^* u^* \\
&= \pm uJuJ^* u^* J^* u^* \\
&= \pm JuJ^* uu^* J^* u^* \\
&= J .
\end{aligned}
\quad (9.55)
$$

Furthermore, by dropping again the symbol π, we have to check that

$$UaU^* = \alpha_u(a) , \quad \forall\, a \in \mathcal{A} , \tag{9.56}$$

$$UDU^* = D_u . \tag{9.57}$$

As for (9.56), for any $a \in \mathcal{A}$ we have,

$$
\begin{aligned}
UaU^* &= uJuJ^*aJu^*J^*u^* \\
&= uJuJ^*Ju^*J^*au^* \quad \text{by 2a. in Definition 6.7.1} \\
&= uau^* \\
&= \alpha_u(a) ,
\end{aligned}
\tag{9.58}
$$

which proves (9.56). Next, by using properties 1b. and 2a., 2b. of Definition 6.7.1 (and their analogues with J and J^* exchanged) the left hand side of (9.57) is given by

$$
\begin{aligned}
UDU^* &= uJuJ^*DJu^*J^*u^* \\
&= uJuDu^*J^*u^* \\
&= uJu(u^*D + [D, u^*])J^*u^* \\
&= uJDJ^*u^* + uJu[D, u^*]J^*u^* \\
&= uDu^* + JJ^*uJu[D, u^*]J^*u^* \\
&= u(u^*D + [D, u^*]) + JuJ^*uJ[D, u^*]J^*u^* \\
&= D + u[D, u^*] + Ju[D, u^*]J^*uJJ^*u^* \\
&= D + u[D, u^*] + Ju[D, u^*]J^* ,
\end{aligned}
\tag{9.59}
$$

and (9.57) is proven.

The operator D_u is interpreted as the product of the perturbation of the 'geometry' given by the operator D, by 'internal gauge degrees of freedom' given by the gauge potential $A = u^*[D, u]$. A general *internal perturbation of the geometry* is provided by

$$D \mapsto D_A = D + A + JAJ^* , \tag{9.60}$$

where A is an arbitrary *gauge potential*, namely an arbitrary Hermitian operator, $A^* = A$, of the form

$$A = \sum_j a_j[D, b_j] , \quad a_j, b_j \in \mathcal{A} . \tag{9.61}$$

Before proceeding, let us observe that for commutative algebras, the internal perturbation $A + JAJ^*$ of the metric in (9.60) vanishes. From what we said after Definition 6.7.1, for commutative algebras one can write $a = Ja^*J^*$ for any $a \in \mathcal{A}$, which amounts to identifying the left multiplicative action by a with the right multiplicative action by Ja^*J^* (always possible if $\mathcal{A}$ is commutative). Furthermore, D is a differential operator of order 1, i.e. $[[D, a], b]] = 0$ for any $a, b \in \mathcal{A}$. Then, with $A = \sum_j a_j[D, b_j]$, $A^* = A$, we get

$$
\begin{aligned}
JAJ^* &= \sum_j Ja_j[D,b_j]J^* = \sum_j Ja_j JJ^*[D,b_j]J^* \\
&= \sum_j a_j^* J[D,b_j]J^* = \sum_j a_j^*[D,Jb_j J^*] \\
&= \sum_j a_j^*[D,b_j^*] = \sum_j [D,b_j^*]a_j^* \\
&= -(a_j \sum_j [D,b_j])^* = -A^* \,,
\end{aligned} \tag{9.62}
$$

and, in turn,

$$
A + JAJ^* = A - A^* = 0 \,. \tag{9.63}
$$

The dynamics of the coupled gravitational and gauge degrees of freedom is governed by a *spectral action principle*. The action is a 'purely geometric' one depending only on the spectrum of the self-adjoint operator D_A [36, 25],

$$
S_B(D,A) = tr_{\mathcal{H}}(\chi(\frac{D_A^2}{\Lambda^2})) \,. \tag{9.64}
$$

Here $tr_{\mathcal{H}}$ is the *usual* trace in the Hilbert space $\mathcal{H}$, Λ is a 'cut off parameter' and χ is a suitable function which cuts off all eigenvalues of D_A^2 larger than Λ^2.

The computation of the action (9.64) is conceptually simple although technically it may be involved. One has just to compute the square of the Dirac operator with Lichnérowicz' formula [10] and the trace with a suitable heat kernel expansion [72], to get an expansion in terms of powers of the parameter Λ. The action (9.64) is interpreted in the framework of Wilson's renormalization group approach to field theory: it gives the *bare* action with *bare coupling constants*. There exists a cut off scale Λ_P which regularizes the action and where the theory is geometric. The renormalized action will have the same form as the bare one with bare parameters replaced by physical parameters [25].
In fact, a full analysis is rather complicated and there are several caveats [63].

In Chap. 10 we shall work out in detail the action for the usual gravitational sector while here we shall indicate how to work it out for a generic gauge field and in particular for the bosonic sector of the standard model. We first proceed with the 'mathematical aspects'.

Proposition 9.3.2. *The spectral action (9.64) is invariant under the gauge action of the inner automorphisms given by*

$$
A \mapsto A^u =: uAu^* + u[D,u^*] \,, \quad \forall\, u \in Inn(\mathcal{A}) \,. \tag{9.65}
$$

Proof. The proof amounts to showing that

$$D_{A^u} = U D_A U^* , \tag{9.66}$$

with U the unitary operator in (9.53), $U = uJuJ^*$. Now, given (9.65), it turns out that

$$
\begin{aligned}
D_{A^u} =: \;& D + A^u + J A^u J^* \\
= \;& D + u[D, u^*] + J[D, u^*]J^* + uAu^* + JuAu^*J^* \\
= \;& D_u + uAu^* + JuAu^*J^*. \tag{9.67}
\end{aligned}
$$

In Proposition 9.3.1 we have already proven that $D_u = UDU^*$, eq.(9.57). To prove the rest, remember that A is of the form $A = \sum_j a_j[D, b_j]$ with $a_j, b_j \in \mathcal{A}$. But, from properties 2a. and 2b. of Definition 6.7.1, it follows that $[A, Jc^*J^*] = 0$, for any $c \in \mathcal{A}$. By using this fact and properties 2a. and 2b. of Definition 6.7.1 (and their analogues with J and J^* exchanged) we have that,

$$
\begin{aligned}
UAU^* =\;& uJuJ^*AJu^*J^*u^* \\
=\;& uJuJ^*Ju^*J^*Au^* \\
=\;& uAu^* . \tag{9.68}
\end{aligned}
$$

$$
\begin{aligned}
U(JAJ^*)U^* =\;& uJuJ^*JAJ^*Ju^*J^*u^* \\
=\;& uJuAu^*J^*u^*JJ^* \\
=\;& uJuAJ^*u^*Ju^*J^* \\
=\;& uJuJ^*u^*JAu^*J^* \\
=\;& JuJ^*uu^*JAu^*J^* \\
=\;& JuAu^*J^* . \tag{9.69}
\end{aligned}
$$

The two previous results together with (9.57) prove eq. (9.66) and, in turn, the proposition.

We spend a few words on the rôle of the outer automorphisms. It turns out that the spectral action (9.64) is *not* invariant under the full $Out(\mathcal{A})$ but rather only under the subgroup $Out(\mathcal{A})^+$ of 'unitarily implementable' elements of $Out(\mathcal{A})$ [37]. Indeed, given any element $\alpha \in Out(\mathcal{A})$, by composition with the representation π of $\mathcal{A}$ in $\mathcal{H}$, one gets an associated representation

$$\pi_\alpha =: \pi \circ \alpha : \mathcal{A} \to \mathcal{B}(\mathcal{H}) . \tag{9.70}$$

The group $Out(\mathcal{A})^+$ is made of elements $\alpha \in Out(\mathcal{A})$ for which there exists a unitary operator U_α on $\mathcal{H}$ such that,

$$\pi_\alpha = U_\alpha \pi U_\alpha^* . \tag{9.71}$$

For the canonical triple on a manifold M, the group $Out(\mathcal{A})^+$ can be identified with $Diff(M)$. We have already mentioned that for $\mathcal{A} = C^\infty(M)$ one has that $Aut(\mathcal{A}) \equiv Out(\mathcal{A}) \simeq Diff(M)$, the last identification being given by pull-back, as in (9.47). We recall now that $\mathcal{H}$ is the Hilbert space $\mathcal{H} = L^2(M, S; d\mu(g))$ of square integrable spinors, the measure $d\mu(g)$ being the canonical Riemannian one associated with the metric g on M. Given an element $\varphi \in Diff(M)$, one defines a unitary operator U_φ on $\mathcal{H}$ by [68],

$$(U_\varphi \psi)(x) =: \left[\frac{\varphi^* d\mu(g)(x)}{d\mu(g)(x)} \right]^{\frac{1}{2}} \psi(\varphi^{-1}(x)) , \quad \forall \, \psi \in \mathcal{H} . \tag{9.72}$$

The unitarity of U_φ follows from the invertibility of φ and the quasi-invariance of the measure $d\mu(g)$ under diffeomorphisms. For any $f \in C^\infty(M)$ it is straightforward to check that (by dropping the symbol π)

$$U_\varphi \, f \, U_\varphi^* = f \circ \varphi^{-1} =: \alpha_\varphi(f) , \tag{9.73}$$

i.e. one reproduces (9.71).

In the usual approach to gauge theories, one constructs connections on a principal bundle $P \to M$ with a finite dimensional Lie group G as structure group. Associated with this bundle there is a sequence of infinite dimensional (Hilbert-Lie) groups which looks remarkably similar to the sequence (9.51) [13, 128],

$$\mathbb{I} \longrightarrow \mathcal{G} \longrightarrow Aut(P) \longrightarrow Diff(M) \longrightarrow \mathbb{I} . \tag{9.74}$$

Here $Aut(P)$ is the group of automorphisms of the total space P, namely diffeomorphisms of P which commute with the action of G, and $\mathcal{G}$ is the subgroup of vertical automorphisms, identifiable with the group of gauge transformations, $\mathcal{G} \simeq C^\infty(M, G)$.

Thus, here is the recipe to construct a spectral gauge theory corresponding to the structure group G or equivalently to the gauge group $\mathcal{G}$ [25]:

1. look for an algebra $\mathcal{A}$ such that $Inn(\mathcal{A}) \simeq \mathcal{G}$;
2. construct a suitable spectral triple 'over' $\mathcal{A}$;
3. compute the spectral action (9.64).

The result will be a gauge theory of the group G coupled with the gravity of the diffeomorphism group $Out(\mathcal{A})$ (with additional extra terms).

For the standard model we have $G = U(1) \times SU(2) \times SU(3)$. It turns out that the relevant spectral triple is the one given in (9.45), (9.46). In fact, as already mentioned in Sect. 9.2, for this triple the structure group would be $U(1) \times SU(2) \times U(3)$; however the computation of $A + JAJ^*$ removes the extra $U(1)$ part from the gauge fields. The associated spectral action has been computed in [25] and in full detail in [76]. The result is the Yang-Mill-Higgs part of the standard model coupled with Einstein gravity plus a cosmological term, a Weyl gravity term and a topological term. Unfortunately the model still suffers from the problems alluded to at the end of Sect. 9.2: namely unrealistic mass relations and an unphysical fermion doubling.

9.4 Fermionic Models

It is also possible to construct the analogue of a gauged Dirac operator by a 'minimal coupling' recipe and to produce an associated action.

If we have a gauge theory on the trivial module $\mathcal{E} = \mathcal{A}$, as in Sec. 8.1, then a gauge potential is just a self-adjoint element $A \in \Omega_D^1 \mathcal{A}$ which transforms under the unitary group $\mathcal{U}(\mathcal{A})$ by (8.5),

$$(A, u) \longrightarrow A^u = uAu^* + u[D, u^*] , \quad \forall\, u \in \mathcal{U}(\mathcal{A}) . \tag{9.75}$$

The following expression is gauge invariant,

$$I_{Dir}(A, \psi) =: \langle \psi, (D + A)\psi \rangle , \quad \forall\, \psi \in Dom(D) \subset \mathcal{H} , \quad A \in \Omega_D^1 \mathcal{A} , \tag{9.76}$$

where the action of the group $\mathcal{U}(\mathcal{A})$ on $\mathcal{H}$ is by restriction of the action of $\mathcal{A}$. Indeed, for any $\psi \in \mathcal{H}$, one has that

$$
\begin{aligned}
(D + A^u)u\psi &= (D + u[D, u^*] + uAu^*)u\psi \\
&= D(u\psi) + u(Du^* - u^*D)(u\psi) + uA\psi \\
&= uDu^*(u\psi) + uA\psi \\
&= u(D + A)\psi ,
\end{aligned}
\tag{9.77}
$$

from which the invariance of (9.76) follows.

The generalization to any finite projective module $\mathcal{E}$ over $\mathcal{A}$ endowed with a Hermitian structure, needs extra care but it is straightforward. In this case one considers the Hilbert space $\mathcal{E} \otimes_{\mathcal{A}} \mathcal{H}$ of 'gauged spinors' introduced in the previous section with the scalar product given in (9.15). The action of the group $End_{\mathcal{A}}(\mathcal{E})$ of endomorphisms of $\mathcal{E}$ extends to an action on $\mathcal{E} \otimes_{\mathcal{A}} \mathcal{H}$ by

$$\phi(\eta \otimes \psi) =: \phi(\eta) \otimes \psi , \quad \forall\, \phi \in End_{\mathcal{A}}(\mathcal{E}) , \quad \eta \otimes \psi \in \mathcal{E} \otimes_{\mathcal{A}} \mathcal{H} . \tag{9.78}$$

In particular, the unitary group $\mathcal{U}(\mathcal{E})$ yields a unitary action on $\mathcal{E} \otimes_{\mathcal{A}} \mathcal{H}$,

$$u(\eta \otimes \psi) =: u(\eta) \otimes \psi , \quad u \in \mathcal{U}(\mathcal{E}) , \quad \eta \otimes \psi \in \mathcal{E} \otimes_{\mathcal{A}} \mathcal{H} , \tag{9.79}$$

since

$$
\begin{aligned}
(u(\eta_1 \otimes \psi_1), u(\eta_2 \otimes \psi_2)) &= (\psi_1, \langle u(\eta_1), u(\eta_2) \rangle \psi_2) \\
&= (\psi_1, \langle \eta_1, \eta_2 \rangle \psi_2) \\
&= (\eta_1 \otimes \psi_1, \eta_2 \otimes \psi_2) ,
\end{aligned}
$$
$$\forall\, u \in \mathcal{U}(\mathcal{E}) , \quad \eta_i \otimes \psi_i \in \mathcal{E} \otimes_{\mathcal{A}} \mathcal{H} \quad , \quad i = 1, 2 . \tag{9.80}$$

If $\nabla : \mathcal{E} \to \mathcal{E} \otimes_{\mathcal{A}} \Omega_D^1 \mathcal{A}$ is a compatible connection on $\mathcal{E}$, the associated 'gauged Dirac operator' D_∇ on the Hilbert space $\mathcal{E} \otimes_{\mathcal{A}} \mathcal{H}$ is defined by

$$D_\nabla(\eta \otimes \psi) = \eta \otimes D\psi + ((\mathbb{I} \otimes \pi)\nabla_{un}\eta)\psi , \quad \eta \in \mathcal{E} , \quad \psi \in \mathcal{H} , \tag{9.81}$$

where ∇_{un} is any universal connection on $\mathcal{E}$ which projects onto ∇.

If $\mathcal{E} = p\mathcal{A}^N$, and $\nabla_{un} = p\delta + \alpha$, then the operator in (9.81) can be written as

$$D_\nabla = pD + \pi(\alpha) , \qquad (9.82)$$

with D acting component-wise on $\mathcal{A}^N \otimes \mathcal{H}$. Since $\pi(\alpha)$ is a self-adjoint operator, from (9.82), we see that D_∇ is a self-adjoint operator on $\mathcal{E} \otimes_\mathcal{A} \mathcal{H}$ with domain $\mathcal{E} \otimes_\mathcal{A} DomD$. Furthermore, since any two universal connections projecting on ∇ differ by $\alpha_1 - \alpha_2 \in ker\pi$, the right-hand side of (9.81) depends only on ∇. Notice that one cannot directly write $(\nabla\eta)\psi$ since $\nabla\eta$ is not an operator on $\mathcal{E} \otimes_\mathcal{A} \mathcal{H}$.

Proposition 9.4.1. *The gauged Dirac action*

$$I_{Dir}(\nabla, \Psi) =: \langle \Psi, D_\nabla\Psi \rangle , \quad \forall \Psi \in \mathcal{E} \otimes_\mathcal{A} DomD , \quad \nabla \in C(\mathcal{E}) , \qquad (9.83)$$

is invariant under the action (9.79) of the unitary group $\mathcal{U}(\mathcal{E})$.

Proof. The proof goes along the same lines of that of (9.77).
For any $\Psi \in \mathcal{E} \otimes_\mathcal{A} \mathcal{H}$, one has that

$$
\begin{aligned}
(pD + \pi(\alpha^u))u\Psi &= (pD + \pi(u\delta u^* + u\alpha u^*))u\Psi \\
&= pD(u\Psi) + u(Du^* - u^*D)(u\Psi) + u\pi(\alpha)\Psi \\
&= pD(u\Psi) + pu(Du^* - u^*D)(u\Psi) + u\pi(\alpha)\Psi \\
&= puDu^*(u\Psi) + u\pi(\alpha)\Psi \\
&= upDu^*(u\Psi) + u\pi(\alpha)\Psi \\
&= u(pD + \pi(\alpha))\Psi ,
\end{aligned}
\qquad (9.84)
$$

which implies the invariance of (9.83).

9.4.1 Fermionic Models on a Two-Point Space

As a very simple example, we shall construct the fermionic Lagrangian (9.76) on the two-point space Y studied in Sects. 6.8 and 9.1.2,

$$I_{Dir}(A, \psi) =: \langle \psi, (D + A)\psi \rangle , \quad \forall \psi \in Dom(D) \subset \mathcal{H} , \quad A \in \Omega_D^1 \mathcal{A} . \quad (9.85)$$

As we have seen in Sect. 6.8, the finite dimensional Hilbert space $\mathcal{H}$ is a direct sum $\mathcal{H} = \mathcal{H}_1 \oplus \mathcal{H}_2$ and the operator D is an off-diagonal matrix

$$D = \begin{bmatrix} 0 & M^* \\ M & 0 \end{bmatrix} , \quad M \in Lin(\mathcal{H}_1, \mathcal{H}_2) . \qquad (9.86)$$

In this simple example $Dom(D) = \mathcal{H}$. On the other hand, a generic gauge potential on the trivial module $\mathcal{E} = \mathcal{A}$ is given by (9.34),

$$A = \begin{bmatrix} 0 & \overline{\Phi}M^* \\ \Phi M & 0 \end{bmatrix} , \quad \Phi \in \mathbb{C} . \qquad (9.87)$$

Summing up, the gauged Dirac operator is the matrix

$$D + A = \begin{bmatrix} 0 & (1 + \overline{\Phi})M^* \\ (1 + \Phi)M & 0 \end{bmatrix} , \tag{9.88}$$

which gives for the action $I_{Dir}(A, \psi)$ a Yukawa-type term coupling the fields $(1 + \Phi)$ and ψ. This action is invariant under the gauge group $\mathcal{U}(\mathcal{E}) = U(1) \times U(1)$.

9.4.2 The Standard Model

Let us now put together the Yang-Mills action (9.14) with the fermionic one in (9.83),

$$\begin{aligned} I(\nabla, \Psi) &= YM(\nabla) + I_{Dir}(\nabla, \Psi) \\ &= \langle F_\nabla, F_\nabla \rangle_2 + \langle \Psi, D_\nabla \Psi \rangle , \qquad \forall \ \nabla \in C(\mathcal{E}) , \\ & \qquad\qquad\qquad\qquad\qquad\qquad \Psi \in \mathcal{E} \otimes_A Dom D . \end{aligned} \tag{9.89}$$

Consider the canonical triple $(\mathcal{A}, \mathcal{H}, D)$ on a Riemannian spin manifold. By taking $\mathcal{E} = \mathcal{A}$, the action (9.89) is just the Euclidean action of massless quantum electrodynamics. If $\mathcal{E} = \mathcal{A}^N$, the action (9.89) is the Yang-Mills action for $U(N)$ coupled with a massless fermion in the fundamental representation of the gauge group $U(N)$ [34].

In [38], the action (9.89) was computed for a product space of a Riemannian spin manifold M with a 'discrete' internal space Y consisting of two points. The result is the full Lagrangian of the standard model. An improved version which uses a real spectral triple and obtained by means of a spectral action along the lines of Sect. 9.3 will be briefly described in the next Section.

9.5 The Fermionic Spectral Action

Consider a real spectral triple $(\mathcal{A}, \mathcal{H}, D, J)$. Recall from Sect. 9.3 the interpretation of gauge degrees of freedom as 'inner fluctuations' of a noncommutative geometry, fluctuations which replace the operator D by $D + A + JAJ^*$, where A is the gauge potential.

The fermionic spectral action is just given by

$$S_F(\psi, A, J) =: \langle \psi, D_A \psi \rangle = \langle \psi, D + A + JAJ^* \rangle \psi \rangle , \tag{9.90}$$

with $\psi \in \mathcal{H}$. The previous action again depends only on the spectral properties of the triple.

By using the $\mathcal{A}$-bimodule structure on $\mathcal{H}$ in (6.77), we get an 'adjoint representation' of the unitary group $\mathcal{U}(\mathcal{A})$ by unitary operators on $\mathcal{H}$,

$$\mathcal{H} \times \mathcal{U}(\mathcal{A}) \ni (\psi, u) \to \psi^u =: u \xi u^* = u J u J^* \ \psi \in \mathcal{H} . \tag{9.91}$$

That this action preserves the scalar product, namely $\langle \psi^u, \psi^u \rangle = \langle \psi, \psi \rangle$, follows from the fact that both u and J act as isometries.

Proposition 9.5.1. *The spectral action (9.90) is invariant under the gauge action of the inner automorphisms given by (9.91) and (9.65),*

$$S_F(\psi^u, A^u, J) = S_F(\psi, A, J) , \quad \forall u \in \mathcal{U}(\mathcal{A}) . \tag{9.92}$$

Proof. By using the result (9.66) $D_{A^u} = U D_A U^*$, with $U = uJuJ^*$, we find

$$
\begin{aligned}
S_F(\psi^u, A^u, J) &= \langle \psi^u, D_{A^u} \psi^u \rangle \\
&= \langle \psi Ju^* J^* u, U D_A U^*)uJuJ^* \psi \rangle \\
&= \langle \psi Ju^* J^* u, uJuJ^* D_A Ju^* J^* uuJuJ^* \psi \rangle \\
&= \langle \psi, D_A \psi \rangle \\
&= S_F(\psi, A, J) .
\end{aligned}
\tag{9.93}
$$

For the spectral triple of the standard model in (9.45), (9.46), (9.41), (9.42), the action (9.90) gives the fermionic sector of the standard model [34, 102]. It is worth stressing that although the noncommutative fermionic multiplet ψ transforms in the adjoint representation (9.91) of the gauge group, the physical fermion fields will transform in the fundamental representation (N) while the antifermions will transform in the conjugate ($\overline{N}$).

10. Gravity Models

We shall describe three possible approaches[1] to the construction of gravity models in noncommutative geometry which, while agreeing for the canonical triple associated with an ordinary manifold (and reproducing the usual Einstein theory), seem to give different answers for more general examples.

As a general remark, we should like to mention that a noncommutative recipe to construct gravity theories (at least the usual Einstein one) has to include the metric as a dynamical variable which is not a priori given. In particular, one should not start with the Hilbert space $\mathcal{H} = L^2(M, S)$ of spinor fields whose scalar product uses a metric on M which, therefore, would play the rôle of a background metric. The beautiful result by Connes [34] which we recall in the following Section goes exactly in the direction of deriving all geometry a posteriori. A possible alternative has been devised in [94].

10.1 Gravity à la Connes-Dixmier-Wodzicki

The first scheme which we use to construct gravity models in noncommutative geometry, and in fact to reconstruct the full geometry out of the algebra $C^\infty(M)$, is based on the Dixmier trace and the Wodzicki residue [36], which we have studied at length in Sects. 6.2 and 6.3.

Proposition 10.1.1. *Suppose we have a smooth compact manifold M without boundary and of dimension n. Let $\mathcal{A} = C^\infty(M)$ and D is just a 'symbol' for the time being. Let $(\mathcal{A}_\pi, D_\pi)$ be a unitary representation of the couple $(\mathcal{A}, D)$ as operators on a Hilbert space $\mathcal{H}_\pi$ endowed with an operator J_π, such that the 'triple' $(\mathcal{A}_\pi, D_\pi, \mathcal{H}_\pi, J_\pi)$ satisfies all the axioms of a real spectral triple given in Sect. 6.4.*
Then,

a) There exists a unique Riemannian metric g_π on M such that the geodesic distance between any two points on M is given by

$$d(p, q) = \sup_{a \in \mathcal{A}}\{|a(p) - a(q)| \mid \|[D_\pi, \pi(a)]\|_{\mathcal{B}(\mathcal{H}_\pi)} \leq 1\}, \quad \forall\, p, q \in M .$$

$$(10.1)$$

---

[1] Two approaches, in fact, since as we shall see the first two are really the same.

b) *The metric g_π depends only on the unitary equivalence class of the representation π. The fibers of the map $\pi \mapsto g_\pi$ from unitary equivalence classes of representations to metrics form a finite collection of affine spaces $\mathcal{A}_\sigma$ parameterized by the spin structures σ on M.*

c) *The action functional given by the Dixmier trace*

$$G(D) = tr_\omega(D^{2-n}) \,, \tag{10.2}$$

is a positive quadratic form with a unique minimum π_σ on each $\mathcal{A}_\sigma$.

d) *The minimum π_σ is the representation of $(\mathcal{A}, D)$ on the Hilbert space of square integrable spinors $L^2(M, S_\sigma)$; $\mathcal{A}_\sigma$ acts by multiplicative operators and D_σ is the Dirac operator of the Levi-Civita connection.*

e) *At the minimum π_σ, the value of $G(D)$ coincides with the Wodzicki residue of D_σ^{2-n} and is proportional to the Einstein-Hilbert action of general relativity*

$$
\begin{aligned}
G(D_\sigma) &= Res_W(D_\sigma^{2-n}) \\
&=: \frac{1}{n(2\pi)^n} \int_{S^*M} tr(\sigma_{-n}(x,\xi))dx d\xi \\
&= c_n \int_M R dx \,,
\end{aligned}
$$

$$c_n = \frac{(2-n)}{12} \frac{2^{[n/2]-n/2}}{(2\pi)^{n/2}} \Gamma(\frac{n}{2}+1)^{-1} \,. \tag{10.3}$$

Here,

$$\sigma_{-n}(x,\xi) = \text{part of order } -n \text{ of the total symbol of } D_\sigma^{2-n} \,, \tag{10.4}$$

R is the scalar curvature of the metric of M and tr is a normalized Clifford trace.

f) *If there is no real structure J, one has to replace spin above by spin^c. The uniqueness of point c) is lost and the minimum of the functional $G(D)$ is reached on a linear subspace of $\mathcal{A}_\sigma$ with σ a fixed spin^c structure. This subspace is parameterized by the $U(1)$ gauge potentials entering in the spin^c Dirac operator. Point d) and c) still hold. In particular the extra terms coming from the $U(1)$ gauge potential drop out of the gravitational action $G(D_\sigma)$.*

Proof. At the moment, a complete proof of this theorem goes beyond our means (and the scope of these notes). We only mention that for $n = 4$ the equality (10.3) was proven by 'brute force' in [87] by means of symbol calculus of pseudodifferential operators. There it was also proven that the results do not depend upon the extra contributions coming from the $U(1)$ gauge potential. In [80], the equality (10.3) was proven in any dimension by realizing that $Res_W(D_\sigma^{2-n})$ is (proportional) to the integral of the second coefficient of the heat kernel expansion of D_σ^2 (see also [2]). It is this fact that relates the previous theorem to the spectral action for gravity as we shall see in the next section.

Finally, the fact that $\mathcal{A}$ is the algebra of smooth functions on a manifold can be recovered a posteriori as well. Connes' axioms allow one to recover the spectrum of $\mathcal{A}$ as a smooth manifold (a smooth submanifold of $\mathbb{R}^N$ for a suitable N) [34].

10.2 Spectral Gravity

In this section we shall compute the spectral action (9.64) described in Sect. 9.3, for the purely gravitational sector. We shall get a sort of 'induced gravity without fermions' together with the corrected sign for the gravitational constant [119, 3].

Consider the canonical triple $(\mathcal{A}, \mathcal{H}, D)$ on a closed n-dimensional Riemannian spin manifold (M, g) which we have described in Sect. 6.5. We recall that $\mathcal{A} = C^\infty(M)$ is the algebra of complex valued smooth functions on M; $\mathcal{H} = L^2(M, S)$ is the Hilbert space of square integrable sections of the irreducible, rank $2^{[n/2]}$ spinor bundle over M; and finally, D is the Dirac operator of the Levi-Civita spin connection.

The action we need to compute is

$$S_G(D, \Lambda) = tr_{\mathcal{H}}(\chi(\frac{D^2}{\Lambda^2})) \ . \tag{10.5}$$

Here $tr_{\mathcal{H}}$ is the usual trace in the Hilbert space $\mathcal{H} = L^2(M, S)$, Λ is a cutoff parameter and χ is a suitable cutoff function which cuts off all the eigenvalues of D^2 which are larger than Λ^2. The parameter Λ has dimension of an inverse length and, as we shall see, determines the scale at which the gravitational action (10.5) departs from the action of general relativity.

As already mentioned, the action (10.5) depends only on the spectrum of D. Before we proceed let us spend some words on the problem of *spectral invariance versus diffeomorphism invariance*. We denote by $spec(M, D)$ the spectrum of the Dirac operator with each eigenvalue repeated according to its multiplicity. Two manifolds M and M' are called *isospectral* if $spec(M, D) = spec(M', D)$.[2] From what has been said, the action (10.5) is a *spectral invariant*. Now, it is well known that one cannot *hear the shape of a drum* [79, 105] (see also [72, 70] and references therein), namely there are manifolds which are isospectral without being isometric (the converse is obviously true). Thus, spectral invariance is stronger than diffeomorphism invariance.

The Lichnérowicz formula (6.53) gives the square of the Dirac operator as

---

[2] In fact, one usually takes the Laplacian instead of the Dirac operator.

$$D^2 = \nabla^S + \frac{1}{4}R \ . \tag{10.6}$$

with R the scalar curvature of the metric, and ∇^S the Laplacian operator lifted to the bundle of spinors,

$$\nabla^S = -g^{\mu\nu}(\nabla^S_\mu \nabla^S_\nu - \Gamma^\rho_{\mu\nu}\nabla^S_\rho) \ , \tag{10.7}$$

and $\Gamma^\rho_{\mu\nu}$ are the Christoffel symbols of the connection.
The heat kernel expansion [72, 25], allows one to express the action (10.5) as an expansion

$$S_G(D,\Lambda) = \sum_{k \geq 0} f_k a_k(D^2/\Lambda^2) \ , \tag{10.8}$$

where the coefficients f_k are given by

$$f_0 = \int_0^\infty \chi(u)u\,du \ ,$$

$$f_2 = \int_0^\infty \chi(u)\,du \ ,$$

$$f_{2(n+2)} = (-1)^n \chi^{(n)}(0) \ , \quad n \geq 0 \ , \tag{10.9}$$

and $\chi^{(n)}$ denotes the n-th derivative of the function χ with respect to its argument.
The Seeley-de Witt coefficients $a_k(D^2/\Lambda^2)$ vanish for odd values of k. The even ones are given as integrals

$$a_k(D^2/\Lambda^2) = \int_M a_k(x; D^2/\Lambda^2)\sqrt{g}\,dx \ . \tag{10.10}$$

The first three coefficients, for even k, are,

$$a_0(x; D^2/\Lambda^2) = (\Lambda^2)^2 \ \frac{1}{(4\pi)^{n/2}} \ tr\mathbb{I}_{2^{[n/2]}} \ ,$$

$$a_2(x; D^2/\Lambda^2) = (\Lambda^2)^1 \ \frac{1}{(4\pi)^{n/2}} \ (-\frac{R}{6} + E) \ tr\mathbb{I}_{2^{[n/2]}}$$

$$a_4(x; D^2/\Lambda^2) = (\Lambda^2)^0 \ \frac{1}{(4\pi)^{n/2}} \frac{1}{360} \ \Big[-12R_{;\mu}^{\ \mu} + 5R^2 - 2R_{\mu\nu}R^{\mu\nu}$$

$$-\frac{7}{4}R_{\mu\nu\rho\sigma}R^{\mu\nu\rho\sigma} - 60RE + 180E^2 + 60E_{;\mu}^{\ \mu} \Big] \ tr\mathbb{I}_{2^{[n/2]}} \ .$$

$$\tag{10.11}$$

Here $R_{\mu\nu\rho\sigma}$ are the components of the Riemann tensor, $R_{\mu\nu}$ the components of the Ricci tensor and R is the scalar curvature. As for E, it is given by $E =: D^2 - \nabla^S = \frac{1}{4}R$. By substituting back into (10.10) and on integrating we obtain

$$a_0(D^2/\Lambda^2) = (\Lambda^2)^2 \, \frac{2^{[n/2]}}{(4\pi)^{n/2}} \int_M \sqrt{g}\,dx \ ,$$

$$a_2(D^2/\Lambda^2) = (\Lambda^2)^1 \, \frac{2^{[n/2]}}{(4\pi)^{n/2}} \frac{1}{12} \int_M \sqrt{g}\,dx \ R \ ,$$

$$a_4(D^2/\Lambda^2) = (\Lambda^2)^0 \, \frac{2^{[n/2]}}{(4\pi)^{n/2}} \frac{1}{360} \int_M \sqrt{g}\,dx \ \Big[3R_{;\mu}^{\ \mu} + \frac{5}{4}R^2$$

$$-2R_{\mu\nu}R^{\mu\nu} - \frac{7}{4}R_{\mu\nu\rho\sigma}R^{\mu\nu\rho\sigma} \Big] \ .$$

$$(10.12)$$

Then, the action (10.5) turns out to be

$$
\begin{aligned}
S_G(D,\Lambda) \ &= \ tr_{\mathcal{H}}(\chi(\frac{D^2}{\Lambda^2})) \\[4pt]
&= \ (\Lambda^2)^2 f_0 \, \frac{2^{[n/2]}}{(4\pi)^{n/2}} \int_M \sqrt{g}\,dx \\[4pt]
&+ \ (\Lambda^2)^1 f_2 \, \frac{2^{[n/2]}}{(4\pi)^{n/2}} \frac{1}{12} \int_M \sqrt{g}\,dx \ R \\[4pt]
&+ \ (\Lambda^2)^0 f_4 \, \frac{2^{[n/2]}}{(4\pi)^{n/2}} \frac{1}{360} \int_M \sqrt{g}\,dx \ \Big[3R_{;\mu}^{\ \mu} + \frac{5}{4}R^2 \\[4pt]
&\qquad\qquad\qquad -2R_{\mu\nu}R^{\mu\nu} - \frac{7}{4}R_{\mu\nu\rho\sigma}R^{\mu\nu\rho\sigma} \Big] \\[4pt]
&+ \ O((\Lambda^2)^{-1}) \ .
\end{aligned}
$$

$$(10.13)$$

Thus we get an action consisting of the Einstein-Hilbert action with a cosmological constant, plus lower order terms in Λ. To get the correct Newton constant in front of the Einstein-Hilbert term, one has to fix the parameter Λ in such a way that $1/\Lambda$ is of the order of the Planck length $L_0 \sim 10^{-33}\,cm$. In turn, this value of Λ produces a *huge* cosmological constant which is a problem for the physical interpretation of the theory if one interprets (10.5) as an approximated gravitational action for slowly varying metrics with small curvature (with respect to the scale Λ). Indeed, the solutions of the equations of motion would have Plank-scale Ricci scalar and, therefore, they would be out of the regime for which the approximation is valid. As we shall see, the cosmological term can be cancelled out.

First of all, we take the function χ to be the characteristic value of the interval $[0, 1]$,

$$\chi(u) = \begin{cases} 1 & \text{for } u \le 1 \ , \\ 0 & \text{for } u \ge 1 \ , \end{cases} \qquad (10.14)$$

possibly 'smoothed out' at $u = 1$. For this choice we get,

$$
\begin{aligned}
f_0 &= 1/2 \ , \quad f_2 = 1 \ , \\
f_4 &= 1 \ , \qquad f_{2(n+2)} = 0 \ , \quad n \ge 1 \ ,
\end{aligned}
\qquad (10.15)
$$

and the action (10.13) becomes

$$
\begin{aligned}
S_G(D, \Lambda) \;=\; & (\Lambda^2)^2 \frac{1}{2} \frac{2^{[n/2]}}{(4\pi)^{n/2}} \int_M \sqrt{g}\, dx \\
& + (\Lambda^2)^1 \frac{2^{[n/2]}}{(4\pi)^{n/2}} \frac{1}{12} \int_M \sqrt{g}\, dx \; R \\
& + (\Lambda^2)^0 \frac{2^{[n/2]}}{(4\pi)^{n/2}} \frac{1}{360} \int_M \sqrt{g}\, dx \left[3R_{;\mu}^{\;\mu} + \frac{5}{4} R^2 \right. \\
& \left. \qquad\qquad\qquad\qquad - 2R_{\mu\nu} R^{\mu\nu} - \frac{7}{4} R_{\mu\nu\rho\sigma} R^{\mu\nu\rho\sigma} \right] .
\end{aligned}
$$

$$(10.16)$$

In [93] the following trick was suggested in order to eliminate the cosmological term: replace the function χ by $\widetilde{\chi}$ defined as

$$
\widetilde{\chi}(u) = \chi(u) - a\chi(bu) , \tag{10.17}
$$

with a, b any two numbers such that $a = b^2$ and $b \geq 0, b \neq 1$. Indeed, one easily finds that,

$$
\begin{aligned}
\widetilde{f}_0 &=: \int_0^\infty \widetilde{\chi}(u) u\, du = (1 - \frac{a}{b^2}) f_0 = 0 , \\
\widetilde{f}_2 &=: \int_0^\infty \widetilde{\chi}(u)\, du = (1 - \frac{a}{b}) f_2 , \\
\widetilde{f}_{2(n+2)} &=: (-1)^n \widetilde{\chi}^{(n)}(0) = (-1)^n (1 - ab^n) \chi^{(n)}(0) , \quad n \geq 0 .
\end{aligned}
$$

$$(10.18)$$

The action (10.5) becomes

$$
\widetilde{S}_G(D, \Lambda) = (1 - \frac{a}{b}) f_2 (\Lambda^2)^1 \frac{2^{[n/2]}}{(4\pi)^{n/2}} \frac{1}{12} \int_M \sqrt{g}\, dx \; R \; + O((\Lambda^2)^0). \tag{10.19}
$$

Low curvature geometries, for which the heat-kernel expansion is valid, are now solutions of the theory.

Summing up, we obtain a theory that approximates pure general relativity at scales which are large when compared with $1/\Lambda$.

We end by mentioning that in [93], in the spirit of spectral gravity, the eigenvalues of the Dirac operator, which are diffeomorphic invariant functions of the geometry[3] and therefore true observables in general relativity, have been taken as a set of variables for an invariant description of the dynamics of the gravitational field. The Poisson brackets of the eigenvalues were computed and determined in terms of the energy-momentum of the eigenspinors and of

---

[3] In fact, the eigenvalues of the Dirac operator are only invariant under the action of diffeomorphisms which preserve the spin structure [15].

the propagator of the linearized Einstein equations. The eigenspinors energy-momenta form the Jacobian of the transformation of coordinates from the metric to the eigenvalues, while the propagator appears as the integral kernel giving the Poisson structure. The equations of motion of the modified action (10.19) are satisfied if the trans Planckian eigenspinors scale linearly with the eigenvalues: this requirement give approximate Einstein equations.

As already mentioned, there exist isospectral manifolds which fail to be isometric. Thus, the eigenvalues of the Dirac operator cannot be used to distinguish among such manifolds (should one really do that from a physical point of view?). A complete analysis of this problem and of its consequences must await another occasion.

10.3 Linear Connections

A different approach to gravity theory, developed in [26, 27], is based on a theory of *linear connections* on an analogue of the cotangent bundle in the noncommutative setting. It turns out that the analogue of the cotangent bundle is more appropriate than the analogue of the tangent bundle. One could define the (analogue) of 'the space of sections of the tangent bundle' as the space of derivations $Der(\mathcal{A})$ of the algebra $\mathcal{A}$. However, in many cases this is not a very useful notion since there are algebras with far too few derivations. Moreover, $Der(\mathcal{A})$ is not an $\mathcal{A}$-module but a module only over the center of $\mathcal{A}$. For models constructed along these lines we refer to [100].

We shall now briefly describe the notion of a linear connection. There are several tricky technical points, mainly related to the Hilbert space closure of space of forms. We ignore them here while referring to [26, 27] for further details.

Suppose then, we have a spectral triple $(\mathcal{A}, \mathcal{H}, D)$ with associated differential calculus $(\Omega_D \mathcal{A}, d)$. The space $\Omega_D^1 \mathcal{A}$ is the analogue of the 'space of sections of the cotangent bundle'. It is naturally a right $\mathcal{A}$-module and we furthermore assume that it is also projective of finite type.

In order to develop 'Riemannian geometry', one needs the 'analogue' of a metric on $\Omega_D^1 \mathcal{A}$. Now, there is a canonical Hermitian structure $\langle \cdot, \cdot \rangle_D$: $\Omega_D^1 \mathcal{A} \times \Omega_D^1 \mathcal{A} \to \mathcal{A}$ which is uniquely determined by the triple $(\mathcal{A}, \mathcal{H}, D)$. It is given by,

$$\langle \alpha, \beta \rangle_D =: P_0(\alpha^* \beta) \in \mathcal{A} , \quad \alpha, \beta \in \Omega_D^1 \mathcal{A} , \qquad (10.20)$$

where P_0 is the orthogonal projector onto $\mathcal{A}$ determined by the scalar product (7.92) as in Sect. 7.3.[4] The map (10.20) satisfies properties (4.19-4.20) which characterize a Hermitian structure. It is also weakly nondegenerate, namely $\langle \alpha, \beta \rangle_D = 0$ for all $\alpha \in \Omega_D^1 \mathcal{A}$ implies that $\beta = 0$. It does not, in general, satisfy the strong nondegeneracy condition expressed in terms of the dual

---

[4] In fact the left hand side of (10.20) is in the completion of $\mathcal{A}$.

module $(\Omega_D^1 \mathcal{A})'$ as in Sect. 4.3. Such a property is assumed to hold. Therefore, if $(\Omega_D^1 \mathcal{A})'$ is the dual module, we assume that the Riemannian structure in (10.20) determines an isomorphism of right $\mathcal{A}$-modules,

$$\Omega_D^1 \mathcal{A} \longrightarrow (\Omega_D^1 \mathcal{A})' , \quad \alpha \mapsto \langle \alpha, \cdot \rangle_D . \tag{10.21}$$

We are now ready to define a linear connection. It is formally the same as in the definition 9.1.1 by taking $\mathcal{E} = \Omega_D^1 \mathcal{A}$.

Definition 10.3.1. *A linear connection on* $\Omega_D^1 \mathcal{A}$ *is a* $\mathbb{C}$-*linear map*

$$\nabla : \Omega_D^1 \mathcal{A} \longrightarrow \Omega_D^1 \mathcal{A} \otimes_\mathcal{A} \Omega_D^1 \mathcal{A} , \tag{10.22}$$

satisfying the Leibniz rule

$$\nabla(\alpha a) = (\nabla \alpha)a + \alpha da , \quad \forall \, \alpha \in \Omega_D^1 \mathcal{A} , \, a \in \mathcal{A} . \tag{10.23}$$

Again, one can extend it to a map $\nabla : \Omega_D^1 \mathcal{A} \otimes_\mathcal{A} \Omega_D^p \mathcal{A} \to \Omega_D^1 \mathcal{A} \otimes_\mathcal{A} \Omega_D^{p+1} \mathcal{A}$ and the *Riemannian curvature* of ∇ is then the $\mathcal{A}$-linear map given by

$$R_\nabla =: \nabla^2 : \Omega_D^1 \mathcal{A} \to \Omega_D^1 \mathcal{A} \otimes_\mathcal{A} \Omega_D^1 \mathcal{A} . \tag{10.24}$$

The connection ∇ is said to be *metric* if it is compatible with the Riemannian structure $\langle \cdot, \cdot \rangle_D$ on $\Omega_D^1 \mathcal{A}$, namely if it satisfies the relation,

$$- \langle \nabla \alpha, \beta \rangle_D + \langle \alpha, \nabla \beta \rangle_D = d \langle \alpha, \beta \rangle_D , \quad \forall \, \alpha, \beta \in \Omega_D^1 \mathcal{A} . \tag{10.25}$$

Next, one defines the *torsion* of the connection ∇ as the map

$$\begin{aligned} T_\nabla : \Omega_D^1 \mathcal{A} &\to \Omega_D^2 \mathcal{A} , \\ T_\nabla &=: d - m \circ \nabla , \end{aligned} \tag{10.26}$$

where $m : \Omega_D^1 \mathcal{A} \otimes_\mathcal{A} \Omega_D^1 \mathcal{A} \to \Omega_D^2 \mathcal{A}$ is just multiplication, $m(\alpha \otimes_\mathcal{A} \beta) = \alpha\beta$. One easily checks (right) $\mathcal{A}$-linearity so that T_∇ is a 'tensor'. For an ordinary manifold with linear connection, the definition(10.26) yields the dual (i.e. the cotangent space version) of the usual definition of torsion.

A connection ∇ on $\Omega_D^1 \mathcal{A}$ is a *Levi-Civita connection* if it is compatible with the Riemannian structure $\langle \cdot, \cdot \rangle_D$ on $\Omega_D^1 \mathcal{A}$ and its torsion vanishes. Contrary to what happens in the ordinary differential geometry, a Levi-Civita connection need not exist for a generic spectral triple or there may exist more than one such connections.

Next, we derive the *Cartan structure equations*. For simplicity, we shall suppose that $\Omega_D^1 \mathcal{A}$ is a free module with a basis $\{E^A, A = 1, \cdots N\}$ so that any element $\alpha \in \Omega_D^1 \mathcal{A}$ can be written as $\alpha = E^A \alpha_A$. The basis is taken to be orthonormal with respect to the Riemannian structure $\langle \cdot, \cdot \rangle_D$,

$$\langle E^A, E^B \rangle_D = \eta^{AB} , \quad \eta^{AB} = diag(1, \cdots, 1) , \quad A, B = 1, \cdots, N . \tag{10.27}$$

A connection ∇ on $\Omega_D^1 \mathcal{A}$ is completely determined by the *connection 1-forms* $\Omega_A{}^B \in \Omega_D^1 \mathcal{A}$ which are defined by,

$$\nabla E^A = E^B \otimes \Omega_B{}^A , \quad A = 1, \ldots, N. \tag{10.28}$$

The components of the torsion $T^A \in \Omega_D^2 \mathcal{A}$ and the curvature $R_A{}^B \in \Omega_D^2 \mathcal{A}$ are defined by

$$T_\nabla(E^A) = T^A ,$$
$$R_\nabla(E^A) = E^B \otimes R_B{}^A , \quad A = 1, \ldots, N. \tag{10.29}$$

By making use of the definitions (10.26) and (10.24) one gets the *structure equations*,

$$T^A = dE^A - E^B \Omega_B{}^A , \quad A = 1, \ldots, N , \tag{10.30}$$
$$R_A{}^B = d\Omega_A{}^B + \Omega_A{}^C \Omega_C{}^B , \quad A, B = 1, \ldots, N. \tag{10.31}$$

The metricity condition (10.25), for the connection 1-forms now reads,

$$- \Omega_C{}^{A*} \eta^{CB} + \eta^{AC} \Omega_C{}^B = 0 . \tag{10.32}$$

As mentioned before, metricity and the vanishing of the torsion do not uniquely fix the connection. Sometimes, one imposes additional constraints by requiring that the connection 1-forms are Hermitian,

$$\Omega_A{}^{B*} = \Omega_A{}^B . \tag{10.33}$$

The components of a connection, its torsion and its Riemannian curvature transform in the 'usual' way under a change of orthonormal basis for $\Omega_D^1 \mathcal{A}$. Consider then a new basis $\{\widetilde{E}^A, A = 1, \cdots N\}$ of $\Omega_D^1 \mathcal{A}$. The relationship between the two bases is given by

$$\widetilde{E}^A = E^B (M^{-1})_B{}^A , \quad E^A = \widetilde{E}^B M_B{}^A , \tag{10.34}$$

with the obvious identities,

$$M_A{}^C (M^{-1})_C{}^B = (M^{-1})_A{}^C M_C{}^B = \delta_A^B , \tag{10.35}$$

which just say that the matrix $M = (M_B{}^A) \in \mathbb{M}_N(\mathcal{A})$ is invertible with inverse given by $M^{-1} = ((M^{-1})_B{}^A)$. By requiring that the new basis be orthonormal with respect to $\langle \cdot, \cdot \rangle_D$ we get,

$$\begin{aligned}
\eta^{AB} &= \langle E^A, E^B \rangle_D \\
&= \langle \widetilde{E}^P M_P{}^A, \widetilde{E}^Q M_Q{}^B \rangle_D \\
&= (M_P{}^A)^* \langle \widetilde{E}^P, \widetilde{E}^Q \rangle_D M_Q{}^B \\
&= (M_P{}^A)^* \eta^{PQ} M_Q{}^B .
\end{aligned} \tag{10.36}$$

From this and (10.35) we obtain the identity

$$(M^{-1})_A{}^B = \eta_{AQ} (M_P{}^Q)^* \eta^{PB} , \tag{10.37}$$

or $M^* = M^{-1}$. From (10.35), we infer that M is a unitary matrix, $MM^* = M^*M = \mathbb{I}$, i.e. it is an element in $\mathcal{U}_N(\mathcal{A})$.

It is now straightforward to find the transformed components of the connection, of its curvature and of its torsion

$$\tilde{\Omega}_A{}^B = M_A{}^P \Omega_P{}^Q (M^{-1})_Q{}^B + M_A{}^P d(M^{-1})_P{}^B , \tag{10.38}$$

$$\tilde{R}_A{}^B = M_A{}^P R_P{}^Q (M^{-1})_Q{}^B , \tag{10.39}$$

$$\tilde{T}^A = T^B (M^{-1})_B{}^A . \tag{10.40}$$

Consider the basis $\{\varepsilon_A, A = 1, \cdots, N\}$ of $(\Omega_D^1 \mathcal{A})'$, dual to the basis $\{E^A\}$,

$$\varepsilon_A(E^B) = \delta_A^B . \tag{10.41}$$

By using the isomorphism (10.21) for the element ε_A, there is an $\hat{\varepsilon}_A \in \Omega_D^1 \mathcal{A}$ determined by

$$\varepsilon_A(\alpha) = \langle \hat{\varepsilon}_A, \alpha \rangle_D , \quad \forall \alpha \in \Omega_D^1 \mathcal{A} , \quad A = 1, \ldots, N. \tag{10.42}$$

One finds that

$$\hat{\varepsilon}_A = E^B \eta_{BA} , \quad A = 1, \ldots, N, \tag{10.43}$$

and, under a change of basis as in (10.34), they transform as

$$\tilde{\hat{\varepsilon}}_A = \hat{\varepsilon}_B (M_A{}^B)^* , \quad A = 1, \ldots, N. \tag{10.44}$$

The *Ricci 1-forms* of the connection ∇ are defined by

$$R_A^\nabla = P_1(R_A{}^B (\hat{\varepsilon}_B)^*) \in \Omega_D^1 \mathcal{A} , \quad A = 1, \cdots, N . \tag{10.45}$$

As for the *scalar curvature*, it is defined by

$$r_\nabla = P_0(E^A R_A^\nabla) = P_0(E^A P_1(R_A{}^B \hat{\varepsilon}_B)^*) \in \mathcal{A} . \tag{10.46}$$

The projectors P_0 and P_1 are again the orthogonal projectors on the space of zero and 1-forms determined by the scalar product (7.92). It is straightforward to check that the scalar curvature does not depend on the particular orthonormal basis of $\Omega_D^1 \mathcal{A}$. Finally, the *Einstein-Hilbert* action is given by

$$I_{HE}(\nabla) = tr_\omega r |D|^{-n} = tr_\omega E^A R_A{}^B \hat{\varepsilon}_B^* |D|^{-n} . \tag{10.47}$$

10.3.1 Usual Einstein Gravity

Let us consider the canonical triple $(\mathcal{A}, \mathcal{H}, D)$ on a closed n-dimensional Riemannian spin manifold (M, g) which we have described in Sect. 6.5. We recall that $\mathcal{A} = C^\infty(M)$ is the algebra of complex valued smooth functions on M; $\mathcal{H} = L^2(M, S)$ is the Hilbert space of square integrable sections of the irreducible spinor bundle over M; and D is the Dirac operator of the Levi-Civita spin connection, which can be written locally as

$$
\begin{aligned}
D &= \gamma^\mu(x)\partial_\mu + \text{ lower order terms} \\
&= \gamma^a e_a^\mu \partial_\mu + \text{ lower order terms} .
\end{aligned}
\tag{10.48}
$$

The 'curved' and 'flat' Dirac matrices are related by

$$
\gamma^\mu(x) = \gamma^a e_a^\mu , \quad \mu = 1, \ldots, n,
\tag{10.49}
$$

and obey the relations

$$
\begin{aligned}
\gamma^\mu(x)\gamma^\nu(x) + \gamma^\nu(x)\gamma^\mu(x) &= -2g^{\mu\nu} , \quad \mu, \nu = 1, \ldots, n, \\
\gamma^a\gamma^b + \gamma^b\gamma^a &= -2\eta^{ab} , \quad a, b = 1, \ldots, n.
\end{aligned}
\tag{10.50}
$$

We shall take the matrices γ^a to be hermitian.

The n-beins e_a^μ relate the components of the curved and flat metric, as usual, by

$$
e_a^\mu g_{\mu\nu} e_b^\nu = \eta_{ab} , \quad e_a^\mu \eta^{ab} e_b^\nu = g^{\mu\nu} .
\tag{10.51}
$$

Finally, we recall that, from the analysis of Sect. 7.2.1, generic elements $\alpha \in \Omega_D^1\mathcal{A}$ and $\beta \in \Omega_D^2\mathcal{A}$ can be written as

$$
\begin{aligned}
\alpha &= \gamma^a \alpha_a = \gamma^\mu \alpha_\mu , \quad \alpha_a = e_a^\mu \alpha_\mu , \\
\beta &= \frac{1}{2}\gamma^{ab}\beta_{ab} = \frac{1}{2}\gamma^{\mu\nu}\beta_{\mu\nu} , \quad \beta_{ab} = e_a^\mu e_b^\nu \beta_{\mu\nu} ,
\end{aligned}
\tag{10.52}
$$

with $\gamma^{ab} = \frac{1}{2}(\gamma^a\gamma^b - \gamma^b\gamma^a)$ and $\gamma^{\mu\nu} = \frac{1}{2}(\gamma^\mu\gamma^\nu - \gamma^\nu\gamma^\mu)$. The module $\Omega_D^1\mathcal{A}$ is projective of finite type and we can take as an orthonormal basis

$$
E^a = \gamma^a , \quad \langle E^a, E^b \rangle = tr\gamma^a\gamma^b = \eta^{ab} , \quad a, b = 1, \ldots, n,
\tag{10.53}
$$

with tr a normalized Clifford trace. Then, the dual basis $\{\varepsilon_a\}$ of $(\Omega_D^1\mathcal{A})'$ is given by,

$$
\varepsilon_a(\alpha) = \alpha_a = e_a^\mu \alpha_\mu ,
\tag{10.54}
$$

and the associated 1-forms $\widehat{\varepsilon}_a$ are found to be

$$
\widehat{\varepsilon}_a = \gamma^a \eta_{ab} .
\tag{10.55}
$$

Hermitian connection 1-forms are of the form

$$
\Omega_a{}^b = \gamma^c \omega_{ca}{}^b = \gamma^\mu \omega_{\mu a}{}^b .
\tag{10.56}
$$

Metricity and the vanishing of torsion read, respectively,

$$\gamma^\mu(\omega_{\mu c}{}^a\eta^{cb} + \eta^{ac}\omega_{\mu c}{}^b) = 0 , \qquad (10.57)$$

$$\gamma^{\mu\nu}(\partial_\mu e_\nu^a - e_\mu^b\omega_{\nu b}{}^a) = 0 . \qquad (10.58)$$

As the sets of matrices $\{\gamma^\mu\}$ and $\{\gamma^{\mu\nu}\}$ are independent, the conditions (10.57) and (10.58) require the vanishing of the terms in parenthesis and, in turn, these just say that the coefficients $\omega_{\mu a}{}^b$ (or equivalently $\omega_{ca}{}^b$) determine the Levi-Civita connection of the metric $g^{\mu\nu}$ [127].
The curvature 2-forms can then be written as

$$R_a{}^b = \frac{1}{2}\gamma^{cd}R_{cda}{}^b , \qquad (10.59)$$

with $R_{cda}{}^b$ the components of the Riemannian tensor of the connection $\omega_{ca}{}^b$. As for the Ricci 1-forms, they are given by

$$R_a = P_1(R_a{}^b\widehat{\varepsilon}_a^*) = \frac{1}{2}\gamma^{cd}\gamma^f R_{cda}{}^b\eta_{fb} . \qquad (10.60)$$

It takes a little algebra to find

$$R_a = -\frac{1}{2}\gamma^c R_{cba}{}^b . \qquad (10.61)$$

The scalar curvature is found to be

$$r =: P_0(\gamma^a R_a) = -\frac{1}{2}P_0(\gamma^a\gamma^c)R_{cba}{}^b = \eta^{ac}R_{cba}{}^b , \qquad (10.62)$$

which is just the usual scalar curvature [127].

10.4 Other Gravity Models

In [26, 27], the action (10.47) was computed for a Connes-Lott space $M \times Y$, i.e. a product of a Riemannian, four-dimensional, spin manifold M with a discrete internal space Y consisting of two points. The Levi-Civita connection on the module of 1-forms depends on a Riemannian metric on M and a real scalar field which determines the distance between the two-sheets. The action (10.47) contains the usual integral of the scalar curvature of the metric on M, a minimal coupling for the scalar field to such a metric, and a kinetic term for the scalar field.
The Wodzicki residue method applied to the same space yields an Einstein-Hilbert action which is the sum of the usual term for the metric of M together with a term proportional to the square of the scalar field. There is no kinetic term for the latter [80].
A somewhat different model of geometry on the Connes-Lott space $M \times Y$ was presented in [95]. The final action is just the Kaluza-Klein action of unified

gravity-electromagnetism and consists of the usual gravity term, a kinetic term for a minimally coupled scalar field and an electromagnetic term.

Several examples of gravity and Kaluza-Klein theories have been constructed by using the bimodule structure of $\mathcal{E} = \Omega^1 \mathcal{A}$ (see [100, 103] and references therein).

11. Quantum Mechanical Models on Noncommutative Lattices

As a very simple example of a quantum mechanical system which can be studied with the techniques of noncommutative geometry on noncommutative lattices, we shall construct the θ-quantization of a particle on a lattice for the circle. We shall do so by constructing an appropriate 'line bundle' with a connection. We refer to [6] and [7] for more details and additional field theoretical examples. In particular, in [7] we derived Wilson's actions for gauge and fermionic fields and analogues of topological and Chern-Simons actions.

The real line $\mathbb{R}^1$ is the universal covering space of the circle S^1, and the fundamental group $\pi_1(S^1) = \mathbb{Z}$ acts on $\mathbb{R}^1$ by translation

$$\mathbb{R}^1 \ni x \to x + N , \ N \in \mathbb{Z} . \tag{11.1}$$

The quotient space of this action is S^1 and the projection : $\mathbb{R}^1 \to S^1$ is given by $\mathbb{R}^1 \ni x \to e^{i2\pi x} \in S^1$.

Now, the domain of a typical Hamiltonian for a particle on S^1 need not consist of functions on S^1. Rather it can be obtained from functions ψ_θ on $\mathbb{R}^1$ transforming under an irreducible representation of $\pi(S^1) = \mathbb{Z}$,

$$\rho_\theta : N \to e^{iN\theta} \tag{11.2}$$

according to

$$\psi_\theta(x + N) = e^{iN\theta} \psi_\theta(x) . \tag{11.3}$$

The domain $D_\theta(H)$ for a typical Hamiltonian H then consists of these ψ_θ restricted to a fundamental domain $0 \leq x \leq 1$ for the action of $\mathbb{Z}$, and subject to a differentiability requirement:

$$D_\theta(H) = \{\psi_\theta : \psi_\theta(1) = e^{i\theta}\psi_\theta(0) \ ; \ \frac{d\psi_\theta(1)}{dx} = e^{i\theta}\frac{d\psi_\theta(0)}{dx}\} . \tag{11.4}$$

In addition, $H\psi_\theta$ must be square integrable for the measure on S^1 used to define the scalar product of wave functions.

One obtains a distinct quantization, called θ-quantization, for each choice of $e^{i\theta}$.

Equivalently, wave functions can be taken to be single-valued functions on S^1 while adding a 'gauge potential' term to the Hamiltonian. To be more

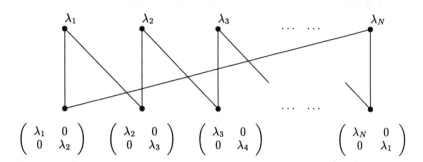

Fig. 11.1. $P_{2N}(S^1)$ for the approximate algebra $\mathcal{C}(\mathcal{A})$

precise, one constructs a line bundle over S^1 with a connection one-form given by $i\theta dx$. If the Hamiltonian with the domain (11.4) is $-d^2/dx^2$, then the Hamiltonian with the domain $D_0(h)$ consisting of single valued wave functions is $-(d/dx + i\theta)^2$.

There are similar quantization possibilities for a noncommutative lattice for the circle as well [6]. One constructs the algebraic analogue of the trivial bundle on the lattice endowed with a gauge connection which is such that the corresponding Laplacian has an approximate spectrum reproducing the 'continuum' one in the limit.

As we have seen in Chap. 3, the algebra $\mathcal{A}$ associated with any non-commutative lattice of the circle is rather complicated and involves infinite dimensional operators on direct sums of infinite dimensional Hilbert spaces. In turn, this algebra $\mathcal{A}$, as it is AF (approximately finite dimensional), can indeed be approximated by algebras of matrices. The simplest approximation is just a commutative algebra $\mathcal{C}(\mathcal{A})$ of the form

$$\mathcal{C}(\mathcal{A}) \simeq \mathbb{C}^N = \{c = (\lambda_1, \lambda_2, \cdots, \lambda_N) , \lambda_i \in \mathbb{C}\} . \qquad (11.5)$$

The algebra (11.5) can produce a noncommutative lattice with $2N$ points by considering a particular class of not necessarily irreducible representations as in Fig. 11.1. In that Figure, the top points correspond to the irreducible one dimensional representations

$$\pi_i : \mathcal{C}(\mathcal{A}) \to \mathbb{C} , \quad c \mapsto \pi_i(c) = \lambda_i , \quad i = 1, \cdots, N . \qquad (11.6)$$

As for the bottom points, they correspond to the reducible two dimensional representations

$$\pi_{i+N} : \mathcal{C}(\mathcal{A}) \to \mathbb{M}_2(\mathbb{C}) , \quad c \mapsto \pi_{i+N}(c) = \begin{pmatrix} \lambda_i & 0 \\ 0 & \lambda_{i+1} \end{pmatrix} , \quad i = 1, \cdots, N ,$$

$$\qquad (11.7)$$

with the additional condition that $N + 1 = 1$. The partial order, or equivalently the topology, is determined by the inclusion of the corresponding kernels as in Chap. 3.

By comparing Fig. 11.1 with the corresponding Fig. 3.18, we see that by trading $\mathcal{A}$ with $\mathcal{C}(\mathcal{A})$, all compact operators have been put to zero. A better approximation is obtained by approximating compact operators with finite dimensional matrices of increasing rank.

The finite projective module of sections $\mathcal{E}$ associated with the 'trivial line bundle' is just $\mathcal{C}(\mathcal{A})$ itself:

$$\mathcal{E} = \mathbb{C}^N = \{\eta = (\mu_1, \mu_2, \cdots, \mu_N), \mu_i \in \mathbb{C}\} . \tag{11.8}$$

The action of $\mathcal{C}(\mathcal{A})$ on $\mathcal{E}$ is simply given by

$$\mathcal{E} \times \mathcal{C}(\mathcal{A}) \to \mathcal{E} , \quad (\eta, c) \mapsto \eta c = (\eta_1 \lambda_1, \eta_2 \lambda_2 \cdots \eta_N \lambda_N) . \tag{11.9}$$

On $\mathcal{E}$ there is a $\mathcal{C}(\mathcal{A})$-valued Hermitian structure $\langle \cdot, \cdot \rangle$,

$$\langle \eta', \eta \rangle := (\eta_1'^* \eta_1, \eta_2'^* \eta_2, \cdots, \eta_N'^* \eta_N) \in \mathcal{C}(\mathcal{A}) . \tag{11.10}$$

Next, we need a K-cycle $(\mathcal{H}, D)$ over $\mathcal{C}(\mathcal{A})$. We take $\mathbb{C}^N$ for $\mathcal{H}$ on which we represent elements of $\mathcal{C}(\mathcal{A})$ as diagonal matrices

$$\mathcal{C}(\mathcal{A}) \ni c \mapsto \mathrm{diag}(\lambda_1, \lambda_2, \ldots \lambda_N) \in \mathcal{B}(\mathbb{C}^N) \simeq \mathbb{M}_N(\mathbb{C}) . \tag{11.11}$$

Elements of $\mathcal{E}$ will be realized in the same manner,

$$\mathcal{E} \ni \eta \mapsto \mathrm{diag}(\eta_1, \eta_2, \ldots \eta_N) \in \mathcal{B}(\mathbb{C}^N) \simeq \mathbb{M}_N(\mathbb{C}) . \tag{11.12}$$

Since our triple $(\mathcal{C}(\mathcal{A}), \mathcal{H}, D)$ will be zero dimensional, the ($\mathbb{C}$-valued) scalar product associated with the Hermitian structure (11.10) will be taken to be

$$(\eta', \eta) = \sum_{j=1}^N \eta_j'^* \eta_j = tr\langle \eta', \eta \rangle , \quad \forall \, \eta', \eta \in \mathcal{E} . \tag{11.13}$$

By identifying $N + j$ with j, we take for the operator D, the $N \times N$ self-adjoint matrix with elements

$$D_{ij} = \frac{1}{\sqrt{2}\epsilon}(m^* \delta_{i+1,j} + m \delta_{i,j+1}) , \quad i,j = 1, \cdots, N , \tag{11.14}$$

where m is any complex number of modulus one: $mm^* = 1$.

As for the connection 1-form ρ on the bundle $\mathcal{E}$, we take it to be the hermitian matrix with elements

$$\rho_{ij} = \frac{1}{\sqrt{2}\epsilon}(\sigma^* m^* \delta_{i+1,j} + \sigma m \delta_{i,j+1}) ,$$

$$\sigma = e^{-i\theta/N} - 1 , \quad i,j = 1, \cdots, N . \tag{11.15}$$

One checks that, modulo junk forms, the curvature of ρ vanishes,

$$d\rho + \rho^2 = 0 . \tag{11.16}$$

It is also possible to prove that ρ is a 'pure gauge' for $\theta = 2\pi k$, with k any integer, that is that there exists a $c \in \mathcal{C}(\mathcal{A})$ such that $\rho = c^{-1}dc$. If $c = \mathrm{diag}(\lambda_1, \lambda_2, \ldots, \lambda_N)$, then any such c will be given by $\lambda_1 = \lambda$, $\lambda_2 = e^{i2\pi k/N}\lambda$, ..., $\lambda_j = e^{i2\pi k(j-1)/N}\lambda$, ..., $\lambda_N = e^{i2\pi k(N-1)/N}\lambda$, with λ not equal to 0 (these properties are the analogues of the properties of the connection $i\theta dx$ in the 'continuum' limit).

The covariant derivative ∇_θ on $\mathcal{E}$, $\nabla_\theta : \mathcal{E} \to \mathcal{E} \otimes_{\mathcal{C}(\mathcal{A})} \Omega^1(\mathcal{C}(\mathcal{A}))$ is then given by

$$\nabla_\theta \eta = [D, \eta] + \rho\eta , \quad \forall \, \eta \in \mathcal{E} . \tag{11.17}$$

In order to define the Laplacian Δ_θ one first introduces a 'dual' operator ∇_θ^* via

$$(\nabla_\theta \eta', \nabla_\theta \eta) = (\eta', \nabla_\theta^* \nabla_\theta \eta) , \quad \forall \, \eta', \eta \in \mathcal{E}. \tag{11.18}$$

The Laplacian Δ_θ on $\mathcal{E}$, $\Delta_\theta : \mathcal{E} \to \mathcal{E}$, can then be defined by

$$\Delta_\theta \eta = -q(\nabla_\theta)^* \nabla_\theta \eta , \quad \forall \, \eta \in \mathcal{E} , \tag{11.19}$$

where q is the orthogonal projector on $\mathcal{E}$ for the scalar product $(\cdot, \cdot)$ in (11.13). This projection operator is readily seen to be given by

$$(qM)_{ij} = M_{ii}\delta_{ij} , \quad \text{no summation on } i , \tag{11.20}$$

with M any element in $\mathbb{M}_N(\mathbb{C})$. Hence, the action of Δ_θ on the element $\eta = (\eta_1, \cdots, \eta_N)$, $\eta_{N+1} = \eta_1$, is explicitly given by

$$
\begin{aligned}
(\Delta_\theta \eta)_{ij} &= -(\nabla_\theta^* \nabla_\theta \eta)_{ii}\delta_{ij} , \\
-(\nabla_\theta^* \nabla_\theta \eta)_{ii} &= \left\{ -[D, [D, \eta]] - 2\rho[D, \eta] - \rho^2\eta \right\}_{ii} \\
&= \frac{1}{\epsilon^2}\left[e^{-i\theta/N}\eta_{i-1} - 2\eta_i + e^{i\theta/N}\eta_{i+1} \right] ; \quad i = 1, 2, \cdots, N .
\end{aligned}
\tag{11.21}
$$

The associated eigenvalue problem

$$\Delta_\theta \eta = \lambda\eta , \tag{11.22}$$

has solutions

$$\lambda = \lambda_k = \frac{2}{\epsilon^2}\left[\cos(k + \frac{\theta}{N}) - 1 \right] , \tag{11.23}$$

$$\eta = \eta^{(k)} = \mathrm{diag}(\eta_1^{(k)}, \eta_2^{(k)}, \cdots, \eta_N^{(k)}) , \quad k = m\frac{2\pi}{N} , \quad m = 1, 2, \cdots, N , \tag{11.24}$$

with each component $\eta_j^{(k)}$ having an expression of the form

$$\eta_j^{(k)} = A^{(k)}e^{ikj} + B^{(k)}e^{-ikj} , \quad A^{(k)}, B^{(k)} \in \mathbb{C} . \tag{11.25}$$

We see that the eigenvalues (11.23) are an approximation to the continuum answers $-4k^2$, $k \in \mathbb{R}$.

A. Appendices

A.1 Basic Notions of Topology

In this appendix we gather together a few fundamental notions of topology and topological spaces while referring to [74, 59] for a more detailed account.

A *topological space* is a set S together with a collection $\tau = \{O_\alpha\}$ of subsets of S, called *open sets*, which satisfy the following axioms

O_1 The union of any number of open sets is an open set.
O_2 The intersection of a finite number of open sets is an open set.
O_3 Both S and the empty set $\emptyset$ are open.

Having a topology it is possible to define the notion of a continuous map. A map $f : (S_1, \tau_1) \to (S_2, \tau_2)$ between two topological spaces is defined to be *continuous* if the inverse image $f^{-1}(O)$ is open in S_1 for any open O in S_2. A continuous map f which is a bijection and such that f^{-1} is continuous as well is called a *homeomorphism*.

Given a topology on a space, one can define the notion of limit point of a subset. A point p is a *limit point* of a subset X of S if every open set containing p contains at least another point of X distinct from p. A subset X of a topological space S is called *closed* if the complement $S \setminus X$ is open. It turns out that the subset X is closed if and only if it contains all its limit points.

The collection $\{C_\alpha\}$ of all closed subsets of a topological space S, satisfy properties which are dual to the corresponding ones for the open sets.

C_1 The intersection of any number of closed sets is a closed set.
C_2 The union of a finite number of closed sets is a closed set.
C_3 Both S and the empty set $\emptyset$ are closed.

One can, then, put a topology on a space by giving a collection of closed sets.

The *closure* $\overline{X}$ of a subset X of a topological space (S, τ) is the intersection of all closed sets containing X. It is evident that $\overline{X}$ is the smallest closed set containing X and that X is closed if and only if $\overline{X} = X$. It turns out that a topology on a set S can be given by means of a *closure operation*. Such an operation is an assignment of a subset $\overline{X}$ of S to any subset X of S, in such a manner that the following *Kuratowski closure axioms* are true,

K_1 $\overline{\emptyset} = \emptyset$.

K_2 $X \subseteq \overline{X}$.

K_3 $\overline{\overline{X}} = \overline{X}$.

K_4 $\overline{X \bigcup Y} = \overline{X} \bigcup \overline{Y}$.

If σ is the family of all subsets X of S for which $\overline{X} = X$ and τ is the family of all complements of members of σ, then τ is a topology for S, and $\overline{X}$ is the τ-closure of X for any subset of S. Clearly, σ is the family of closed sets.

A topological space S is said to be a T_0-*space* if: given any two points of S, at least one of them is contained in an open set not containing the other. This can also be stated by saying that for any pair of points, at least one of the points is not a limit point of the other. In such a space, there may be sets consisting of a single point which are not closed.

A topological space S is said to be a T_1-*space* if: given any two points of S, each of them lies in an open set not containing the other. This requirement implies that each point (and then, by C_2 above, every finite set) is closed. This is often taken as a definition of a T_1-space.

A topological space S is said to be a T_2-*space* or a *Hausdorff space* if: given any two points of S, there are *disjoint* open sets each containing one of the two points but not both.

It is clear that the previous conditions are in an increasing order of strength in the sense that being T_2 implies being T_1 and being T_1 implies being T_0 (a space which is T_2 is T_1 and a space which is T_1 is T_0.).

A family $\mathcal{U}$ of sets is a *cover* of a (topological) space if $S = \bigcup \{X, X \in \mathcal{U}\}$. The family is an *open cover* of S if every member of $\mathcal{U}$ is an open set. The family is a *finite cover* if the number of members of $\mathcal{U}$ is finite. It is a *locally finite* cover if and only if every $x \in S$ has a neighborhood that intersects only a finite number of members of the family.

A topological space S is called *compact* if every open cover of S has a finite subcover of S. A topological space S is called *locally compact* if any point of S has at least one compact neighborhood. A compact space is automatically locally compact. If S is a locally compact space which is also Hausdorff, then the family of closed compact neighborhoods of any point is a basis for its neighborhood system.

The *support* of a real or complex valued function f on a topological space S is the closure of the set $K_f = \{x \in S \mid f(x) \neq 0\}$. The function f is said to have *compact support* if K_f is compact. The collection of all continuous functions on S whose support is compact is denoted by $C_c(S)$.

A real or complex valued function f on a locally compact Hausdorff space S is said to *vanish at infinity* if for every $\epsilon > 0$ there exists a compact set $K \subset S$ such that $|f(x)| < \epsilon$ for all $x \notin K$. The collection of all continuous functions on S which vanish at infinity is denoted by $C_0(S)$. Clearly $C_c(S) \subset C_0(S)$, and the two classes coincide if S is compact. Furthermore, one can prove that $C_0(S)$ is the completion of $C_c(S)$ relative to the supremum norm (2.9) described in Example 2.1.1 [118].

A continuous map between two locally compact Hausdorff spaces $f : S_1 \to S_2$ is called *proper* if and only if for any compact subset K of S_2, the inverse image $f^{-1}(K)$ is a compact subset of S_1.

A space which contains a dense subset is called *separable*. A topological space which has a countable basis of open sets is called *second-countable* (or *completely separable*).

A topological space S is called *connected* if it is not the union of two disjoint, nonempty open sets. Equivalently, if the only sets in S that are both open and closed are S and the empty set then S is connected. A subset C of the topological space S is called a *component* of S, provided that C is connected and maximal, namely it is not a proper subset of another connected set in S. One can prove that any point of S lies in a component. A topological space is called *totally disconnected* if the (connected) component of each point consists only of the point itself. The *Cantor set* is a totally disconnected space. In fact, any totally disconnected, second countable, compact Hausdorff space is homeomorphic to a subset of the Cantor set.

If τ_1 and τ_2 are two topologies on the space S, one says that τ_1 is *coarser* than τ_2 (or that τ_2 is *finer* than τ_1) if and only if $\tau_1 \subset \tau_2$, namely if and only if every subset of S which is open in τ_1 is also open in τ_2. Given two topologies on the space S it may happen that neither of them is coarser (or finer) than the other. The set of all possible topologies on the same space is a partially ordered set whose *coarsest* element is the topology in which only $\emptyset$ and S are open, while the *finest* element is the topology in which all subsets of S are open (this topology is called the discrete topology).

A.2 The Gel'fand-Naimark-Segal Construction

A *state* on the C^*-algebra $\mathcal{A}$ is a linear functional

$$\phi : \mathcal{A} \longrightarrow \mathbb{C} , \tag{A.1}$$

which is positive and of norm one, i.e.

$$\phi(a^*a) \geq 0 , \quad \forall \, a \in \mathcal{A} ,$$
$$||\phi|| = 1 . \tag{A.2}$$

Here the norm of ϕ is defined as usual by $||\phi|| = sup\{|\phi(a)| \mid ||a|| \leq 1\}$. If $\mathcal{A}$ has a unit (we always assume this is the case), positivity implies that

$$||\phi|| = \phi(\mathbb{I}) = 1 . \tag{A.3}$$

The set $\mathcal{S}(\mathcal{A})$ of all states of $\mathcal{A}$ is clearly a convex space, since the convex combination $\lambda\phi_1 + (1 - \lambda)\phi_2 \in \mathcal{S}(\mathcal{A})$, for any $\phi_1, \phi_2 \in \mathcal{S}(\mathcal{A})$ and $0 \leq \lambda \leq 1$. Elements at the boundary of $\mathcal{S}(\mathcal{A})$ are called *pure states*, namely, a state ϕ is called pure if it cannot be written as the convex combination of (two) other states. The space of pure states is denoted by $\mathcal{PS}(\mathcal{A})$. If the algebra $\mathcal{A}$ is abelian, a pure state is the same as a character and the space $\mathcal{PS}(\mathcal{A})$ is just the space $\widehat{\mathcal{A}}$ of characters of $\mathcal{A}$ which, when endowed with the Gel'fand topology, is a Hausdorff (locally compact) topological space.

With each state $\phi \in \mathcal{S}(\mathcal{A})$ there is associated a representation $(\mathcal{H}_\phi, \pi_\phi)$ of $\mathcal{A}$, called the Gel'fand-Naimark-Segal (GNS) representation. The procedure to construct such a representation is also called the GNS construction which we shall now briefly describe [43, 108].

Suppose then that we are given a state $\phi \in \mathcal{S}(\mathcal{A})$ and consider the space

$$\mathcal{N}_\phi = \{a \in \mathcal{A} \mid \phi(a^*a) = 0\} . \tag{A.4}$$

By using the fact that $\phi(a^*b^*ba) \leq ||b||^2 \phi(a^*a)$, one infers that $\mathcal{N}_\phi$ is a closed (left) ideal of $\mathcal{A}$. The space $\mathcal{A}/\mathcal{N}_\phi$ of equivalence classes is made into a pre-Hilbert space by defining a scalar product by

$$\mathcal{A}/\mathcal{N}_\phi \times \mathcal{A}/\mathcal{N}_\phi \longrightarrow \mathbb{C} , \quad (a + \mathcal{N}_\phi, b + \mathcal{N}_\phi) \mapsto \phi(a^*b) . \tag{A.5}$$

This scalar product is clearly independent of the representatives in the equivalence classes.

The Hilbert space $\mathcal{H}_\phi$ completion of $\mathcal{A}/\mathcal{N}_\phi$ is the space of the representation. Then, to any $a \in \mathcal{A}$ one associates an operator $\pi(a) \in \mathcal{B}(\mathcal{A}/\mathcal{N}_\phi)$ by

$$\pi(a)(b + \mathcal{N}_\phi) =: ab + \mathcal{N}_\phi . \tag{A.6}$$

Again, this action does not depend on the representative. By using the fact that $||\pi(a)(b + \mathcal{N}_\phi)||^2 = \phi(b^*a^*ab) \leq ||a||^2 \phi(b^*b) = ||b + \mathcal{N}_\phi||^2$, one gets $||\pi(a)|| \leq ||a||$ and in turn, that $\pi(a) \in \mathcal{B}(\mathcal{A}/\mathcal{N}_\phi)$. There is a unique extension of $\pi(a)$ to an operator $\pi_\phi(a) \in \mathcal{B}(\mathcal{H}_\phi)$. Finally, one easily checks the algebraic properties $\pi_\phi(a_1 a_2) = \pi_\phi(a_1)\pi_\phi(a_2)$ and $\pi_\phi(a^*) = (\pi_\phi(a))^*$ and one obtains a $*$-morphism (a representation)

$$\pi_\phi : \mathcal{A} \longrightarrow \mathcal{B}(\mathcal{H}_\phi) , \quad a \mapsto \pi_\phi(a) . \tag{A.7}$$

It turns out that any state ϕ is a *vector state*. This means that there exists a vector $\xi_\phi \in \mathcal{H}_\phi$ with the property,

$$(\xi_\phi, \pi_\phi(a)\xi_\phi) = \phi(a) , \forall\, a \in \mathcal{A} . \tag{A.8}$$

Such a vector is defined by

$$\xi_\phi =: [\mathbb{I}] = \mathbb{I} + \mathcal{N}_\phi , \tag{A.9}$$

and is readily seen to verify (A.8). Furthermore, the set $\{\pi_\phi(a)\xi_\phi \mid a \in \mathcal{A}\}$ is just the dense set $\mathcal{A}/\mathcal{N}_\phi$ of equivalence classes. This fact is stated by saying that the vector ξ_ϕ is a *cyclic vector* for the representation $(\mathcal{H}_\phi, \pi_\phi)$. By construction, and by (A.3), a cyclic vector is of norm one, $||\xi_\phi||^2 = ||\phi|| = 1$.

The cyclic representation $(\mathcal{H}_\phi, \pi_\phi, \xi_\phi)$ is unique up to unitary equivalence. If $(\mathcal{H}'_\phi, \pi'_\phi, \xi'_\phi)$ is another cyclic representation such that $(\xi'_\phi, \pi'_\phi(a)\xi'_\phi) = \phi(a)$, for all $a \in \mathcal{A}$, then there exists a unitary operator $U : \mathcal{H}_\phi \to \mathcal{H}'_\phi$ such that

$$U^{-1}\pi'_\phi(a)U = \pi_\phi(a) , \quad \forall\, a \in \mathcal{A} ,$$
$$U\xi_\phi = \xi'_\phi . \tag{A.10}$$

The operator U is defined by $U\pi_\phi(a)\xi_\phi = \pi'_\phi(a)\xi'_\phi$ for any $a \in \mathcal{A}$. Then, the properties of the state ϕ ensure that U is well defined and preserves the scalar product.

It is easy to see that the representation $(\mathcal{H}_\phi, \xi_\phi)$ is irreducible if and only if every non zero vector $\xi \in \mathcal{H}_\phi$ is cyclic so that there are no nontrivial invariant subspaces. It is somewhat surprising that this happens exactly when the state ϕ is pure [43].

Proposition A.2.1. *Let $\mathcal{A}$ be a C^*-algebra. Then,*

1. *A state ϕ on $\mathcal{A}$ is pure if and only if the associated GNS representation $(\mathcal{H}_\phi, \pi_\gamma)$ is irreducible.*
2. *Given a pure state ϕ on $\mathcal{A}$ there is a canonical bijection between rays in the associated Hilbert $\mathcal{H}_\phi$ and the equivalence class of ϕ,*

$$C_\phi = \{\psi \text{ pure state on } \mathcal{A} \mid \pi_\psi \text{ equivalent to } \pi_\phi\} .$$

The bijection of point 2. of the previous preposition is explicitly given by associating with any $\xi \in \mathcal{H}_\phi$, $||\xi|| = 1$, the state on $\mathcal{A}$ given by

$$\psi(a) = (\xi, \pi_\phi(a)\xi) , \quad \forall\, a \in \mathcal{A} , \tag{A.11}$$

which is seen to be pure. As was previously noted, the representation $(\mathcal{H}_\phi, \pi_\phi)$ is irreducible and each vector of $\mathcal{H}_\phi$ is cyclic. This, in turn, implies that the representation associated with the state ψ is equivalent to $(\mathcal{H}_\phi, \pi_\phi)$.

As a simple example, we consider the algebra $\mathbf{M}_2(\mathbb{C})$ with the two pure states constructed in Sect. 2.3,

$$\phi_1\left(\begin{bmatrix} a_{11} & a_{12} \\ a_{21} & a_{22} \end{bmatrix}\right) = a_{11}, \quad \phi_2\left(\begin{bmatrix} a_{11} & a_{12} \\ a_{21} & a_{22} \end{bmatrix}\right) = a_{22}. \tag{A.12}$$

As we mentioned before, the corresponding representations are equivalent. We shall show that they are both equivalent to the defining two dimensional one.

The ideals of elements of 'vanishing norm' of the states ϕ_1, ϕ_2 are, respectively,

$$\mathcal{N}_1 = \left\{\begin{bmatrix} 0 & a_{12} \\ 0 & a_{22} \end{bmatrix}\right\}, \quad \mathcal{N}_2 = \left\{\begin{bmatrix} a_{11} & 0 \\ a_{21} & 0 \end{bmatrix}\right\}. \tag{A.13}$$

The associated Hilbert spaces are then found to be

$$\mathcal{H}_1 = \left\{\begin{bmatrix} x_1 & 0 \\ x_2 & 0 \end{bmatrix}\right\} \simeq \mathbb{C}^2 = \left\{X = \begin{pmatrix} x_1 \\ x_2 \end{pmatrix}\right\},$$
$$\langle X, X' \rangle = x_1^* x_1' + x_2^* x_2'.$$

$$\mathcal{H}_2 = \left\{\begin{bmatrix} 0 & y_1 \\ 0 & y_2 \end{bmatrix}\right\} \simeq \mathbb{C}^2 = \left\{X = \begin{pmatrix} y_1 \\ y_2 \end{pmatrix}\right\},$$
$$\langle Y, Y' \rangle = y_1^* y_1' + y_2^* y_2'. \tag{A.14}$$

As for the action of any element $A \in \mathbf{M}_2(\mathbb{C})$ on $\mathcal{H}_1$ and $\mathcal{H}_2$, we get

$$\pi_1(A)\begin{bmatrix} x_1 & 0 \\ x_2 & 0 \end{bmatrix} = \begin{bmatrix} a_{11}x_1 + a_{12}x_2 & 0 \\ a_{21}x_1 + a_{22}x_2 & 0 \end{bmatrix} \equiv A\begin{pmatrix} x_1 \\ x_2 \end{pmatrix},$$
$$\pi_2(A)\begin{bmatrix} 0 & y_1 \\ 0 & y_2 \end{bmatrix} = \begin{bmatrix} 0 & a_{11}y_1 + a_{12}y_2 \\ 0 & a_{21}y_1 + a_{22}y_2 \end{bmatrix} \equiv A\begin{pmatrix} y_1 \\ y_2 \end{pmatrix}. \tag{A.15}$$

The two cyclic vectors are given by

$$\xi_1 = \begin{pmatrix} 1 \\ 0 \end{pmatrix}, \quad \xi_2 = \begin{pmatrix} 0 \\ 1 \end{pmatrix}. \tag{A.16}$$

The equivalence of the two representations is provided by the off-diagonal matrix

$$U = \begin{bmatrix} 0 & 1 \\ 1 & 0 \end{bmatrix}, \tag{A.17}$$

which interchanges 1 and 2 , $U\xi_1 = \xi_2$. Indeed, by using the fact that for an irreducible representation any non vanishing vector is cyclic, from (A.15) we see that the two representations can indeed be identified.

A.3 Hilbert Modules

The theory of Hilbert modules is a generalization of the theory of Hilbert spaces and it is the natural framework for the study of modules over a C^*-algebra $\mathcal{A}$ endowed with a Hermitian $\mathcal{A}$-valued inner product. Hilbert modules have been (and are) used in a variety of applications, notably for the study of strong Morita equivalence. The subject started with the works [113] and [109]. We refer to [131] for a very nice introduction while here we simply report on the fundamentals of the theory. Throughout this appendix, $\mathcal{A}$ will be a C^*-algebra (almost always with unit) and its norm will be denoted simply by $\| \cdot \|$.

Definition A.3.1. *A right pre-Hilbert module over $\mathcal{A}$ is a right $\mathcal{A}$-module $\mathcal{E}$ endowed with an $\mathcal{A}$-valued Hermitian structure, namely a sesquilinear form $\langle \, , \, \rangle_{\mathcal{A}} : \mathcal{E} \times \mathcal{E} \to \mathcal{A}$, which is conjugate linear in the first variable and such that*

$$\langle \eta_1, \eta_2 a \rangle_{\mathcal{A}} = \langle \eta_1, \eta_2 \rangle_{\mathcal{A}} \, a \, , \tag{A.18}$$

$$\langle \eta_1, \eta_2 \rangle_{\mathcal{A}}^* = \langle \eta_2, \eta_1 \rangle_{\mathcal{A}} \, , \tag{A.19}$$

$$\langle \eta, \eta \rangle_{\mathcal{A}} \geq 0 \, , \quad \langle \eta, \eta \rangle_{\mathcal{A}} = 0 \Leftrightarrow \eta = 0 \, , \tag{A.20}$$

for all $\eta_1, \eta_2, \eta \in \mathcal{E}$, $a \in \mathcal{A}$.

By the property (A.20) in the previous definition the element $\langle \eta, \eta \rangle_{\mathcal{A}}$ is self-adjoint. As in ordinary Hilbert spaces, the property (A.20) provides a *generalized Cauchy-Schwartz inequality*

$$\langle \eta, \xi \rangle_{\mathcal{A}}^* \langle \eta, \xi \rangle_{\mathcal{A}} \leq \| \langle \eta, \eta \rangle_{\mathcal{A}} \| \langle \xi, \xi \rangle_{\mathcal{A}} \, , \quad \forall \, \eta, \xi \in \mathcal{E} \, , \tag{A.21}$$

which in turns, implies

$$\| \langle \eta, \xi \rangle_{\mathcal{A}} \|^2 \leq \| \langle \eta, \eta \rangle_{\mathcal{A}} \| \| \langle \xi, \xi \rangle_{\mathcal{A}} \| \, , \quad \forall \, \eta, \xi \in \mathcal{E} \, , \tag{A.22}$$

By using these properties and the norm $\| \cdot \|$ in $\mathcal{A}$ one defines a norm in $\mathcal{E}$.

Definition A.3.2. *The norm of any element $\eta \in \mathcal{E}$ is defined by*

$$\| \eta \|_{\mathcal{A}} =: \sqrt{\| \langle \eta, \eta \rangle \|} \, . \tag{A.23}$$

One can prove that $\| \cdot \|_{\mathcal{A}}$ satisfies all the properties (2.5) of a norm.

Definition A.3.3. *A right Hilbert module over $\mathcal{A}$ is a right pre-Hilbert module $\mathcal{E}$ which is complete with respect to the norm $\| \cdot \|_{\mathcal{A}}$.*

By completion any right pre-Hilbert module will give a right Hilbert module. It is clear that Hilbert modules over $\mathbb{C}$ are ordinary Hilbert spaces.

A *left* (pre-)Hilbert module structure on a left $\mathcal{A}$-module $\mathcal{E}$ is provided by an $\mathcal{A}$-valued Hermitian structure $\langle \, , \, \rangle_{\mathcal{A}}$ on $\mathcal{E}$ which is conjugate linear in the second variable and with condition (A.18) replaced by

$$\langle a\eta_1, \eta_2 \rangle_A = a \langle \eta_1, \eta_2 \rangle_A \ , \quad \forall \ \eta_1, \eta_2, \in \mathcal{E}, \ a \in \mathcal{A} \ . \tag{A.24}$$

In what follows, unless otherwise stated, by a Hilbert module we shall mean a right one. It is straightforward to pass to equivalent statements concerning left modules.

Given any Hilbert module $\mathcal{E}$ over $\mathcal{A}$, the closure of the linear span of $\{\langle \eta_1, \eta_2 \rangle_A \ , \ \eta_1, \eta_2 \in \mathcal{E}\}$ is an ideal in $\mathcal{A}$. If this ideal is the whole of $\mathcal{A}$ the module $\mathcal{E}$ is called a *full Hilbert module.*[1]

It is worth noticing that, contrary to what happens in an ordinary Hilbert space, the Pythagoras equality is not valid in a generic Hilbert module $\mathcal{E}$. If η_1 and η_2 are any two orthogonal elements in $\mathcal{A}$, i.e. $\langle \eta_1, \eta_2 \rangle_A = 0$, in general one has that $||\eta_1 + \eta_2||_A^2 \neq ||\eta_1||_A^2 + ||\eta_2||_A^2$. Indeed, properties of the norm only guarantee that $||\eta_1 + \eta_2||_A^2 \leq ||\eta_1||_A^2 + ||\eta_2||_A^2$.

An 'operator' on a Hilbert module need not admit an adjoint.

Definition A.3.4. *Let $\mathcal{E}$ be a Hilbert module over the C^*-algebra $\mathcal{A}$. A continuous $\mathcal{A}$-linear map $T : \mathcal{E} \to \mathcal{E}$ is said to be* adjointable *if there exists a map $T^* : \mathcal{E} \to \mathcal{E}$ such that*

$$\langle T^*\eta_1, \eta_2 \rangle_A = \langle \eta_1, T\eta_2 \rangle_A \ , \quad \forall \ \eta_1, \eta_2 \in \mathcal{E} \ . \tag{A.25}$$

The map T^* is called the adjoint of T. We shall denote by $End_A(\mathcal{E})$ the collection of all continuous $\mathcal{A}$-linear adjointable maps. Elements of $End_A(\mathcal{E})$ will also be called endomorphisms of $\mathcal{E}$.

One can prove that if $T \in End_A(\mathcal{E})$, then also its adjoint $T^* \in End_A(\mathcal{E})$ with $(T^*)^* = T$. If both T and S are in $End_A(\mathcal{E})$, then $TS \in End_A(\mathcal{E})$ with $(TS)^* = S^*T^*$. Finally, the space $End_A(\mathcal{E})$, endowed with this involution and with the operator norm

$$||T|| =: sup\{||T\eta||_A \ | \ ||\eta||_A \leq 1\} \ , \tag{A.26}$$

becomes a C^*-algebra of bounded operators due to the inequality $\langle T\eta, T\eta \rangle_A \leq ||T||^2 \langle \eta, \eta \rangle_A$. Indeed, $End_A(\mathcal{A})$ is complete if $\mathcal{E}$ is.

There are also the analogues of *compact endomorphisms* which are obtained as usual from 'endomorphisms of finite rank'. For any $\eta_1, \eta_2 \in \mathcal{E}$ an endomorphism $|\eta_1\rangle \langle \eta_2|$ is defined by

$$|\eta_1\rangle \langle \eta_2| (\xi) =: \eta_1 \langle \eta_2, \xi \rangle_A \ , \quad \forall \xi \in \mathcal{E} \ . \tag{A.27}$$

Its adjoint is just given by

$$(|\eta_1\rangle \langle \eta_2|)^* = |\eta_2\rangle \langle \eta_1| \ , \quad \forall \ \eta_1, \eta_2 \in \mathcal{E} \ . \tag{A.28}$$

One can check that

$$|| \ |\eta_1\rangle \langle \eta_2| \ ||_A \leq ||\eta_1||_A ||\eta_2||_A \ , \quad \forall \xi \in \mathcal{E} \ . \tag{A.29}$$

---

[1] Rieffel [113] calls it an $\mathcal{A}$-*rigged space.*

Furthermore, for any $T \in End_A(\mathcal{E})$ and any $\eta_1, \eta_2, \xi_1, \xi_2 \in \mathcal{E}$, one has the expected composition rules

$$T \circ |\eta_1\rangle \langle \eta_2| = |T\eta_1\rangle \langle \eta_2| , \tag{A.30}$$

$$|\eta_1\rangle \langle \eta_2| \circ T = |\eta_1\rangle \langle T^*\eta_2| , \tag{A.31}$$

$$|\eta_1\rangle \langle \eta_2| \circ |\xi_1\rangle \langle \xi_2| = |\eta_1 \langle \eta_2, \xi_1\rangle_A\rangle \langle \xi_2| = |\eta_1\rangle \langle \langle \eta_2, \xi_1\rangle_A \xi_2| . \tag{A.32}$$

From these rules, we get that the linear span of the endomorphisms of the form (A.27) is a self-adjoint two-sided ideal in $End_A(\mathcal{E})$. The norm closure in $End_A(\mathcal{E})$ of this two-sided ideal is denoted by $End_A^0(\mathcal{E})$; its elements are called *compact endomorphisms* of $\mathcal{E}$.

Example A.3.1. The Hilbert module $\mathcal{A}$.
The C^*-algebra $\mathcal{A}$ can be made into a (full) Hilbert Module by considering it as a *right* module over itself together with the following Hermitian structure

$$\langle \, , \, \rangle_A : \mathcal{E} \times \mathcal{E} \to \mathcal{A} , \quad \langle a, b\rangle_A =: a^* b , \quad \forall \, a, b \in \mathcal{A} . \tag{A.33}$$

The corresponding norm coincides with the norm of $\mathcal{A}$ since from the norm property (2.8), $\|a\|_A = \sqrt{\|\langle a, a\rangle_A\|} = \sqrt{\|a^* a\|} = \sqrt{\|a\|^2} = \|a\|$. Thus, $\mathcal{A}$ is complete also as a Hilbert module. Furthermore, as the algebra $\mathcal{A}$ is unital, one finds that $End_A(\mathcal{A}) \simeq End_A^0(\mathcal{A}) \simeq \mathcal{A}$, with the latter acting as multiplicative operators on the *left* on itself. In particular, the isometric isomorphism $End_A^0(\mathcal{A}) \simeq \mathcal{A}$ is given by

$$End_A^0(\mathcal{A}) \ni \sum_k \lambda_k |a_k\rangle \langle \beta_k| \mapsto \sum_k \lambda_k a_k \beta_k^* , \quad \forall \, \lambda_k \in \mathbb{C} , \; a_k, b_k \in \mathcal{A} . \tag{A.34}$$

Example A.3.2. The Hilbert module $\mathcal{A}^N$.
Let $\mathcal{A}^N = \mathcal{A} \times \cdots \times \mathcal{A}$ be the direct sum of N copies of $\mathcal{A}$. It is promoted to a full Hilbert module over $\mathcal{A}$ with module action and Hermitian product given by

$$(a_1, \cdots, a_N)a =: (a_1 a, \cdots, a_N a) , \tag{A.35}$$

$$\langle (a_1, \cdots, a_N), (b_1, \cdots, b_N)\rangle_A =: \sum_{k=1}^{n} a_k^* b_k , \tag{A.36}$$

for all $a, a_k, b_k \in \mathcal{A}$. The corresponding norm is

$$\|(a_1, \cdots, a_N)\|_A =: \|\sum_{k=1}^{n} a_k^* a_k\| . \tag{A.37}$$

That $\mathcal{A}^N$ is complete in this norm is a consequence of the completeness of $\mathcal{A}$ with respect to its norm. Indeed, if $(a_1^\alpha, \cdots, a_N^\alpha)_{\alpha \in \mathbb{N}}$ is a Cauchy sequence in $\mathcal{A}^N$, then, for each component, $(a_k^\alpha)_{\alpha \in \mathbb{N}}$ is a Cauchy sequence in $\mathcal{A}$. The limit of $(a_1^\alpha, \cdots, a_N^\alpha)_{\alpha \in \mathbb{N}}$ in $\mathcal{A}^N$ is just the collection of the limits from each component.

Since $\mathcal{A}$ is taken to be unital, the unit vectors $\{e_k\}$ of $\mathbb{C}^N$ form an orthonormal basis for $\mathcal{A}^N$ and each element of $\mathcal{A}^N$ can be written uniquely as $(a_1, \cdots, a_N) = \sum_{k=1}^N e_k a_k$ giving an identification $\mathcal{A}^N \simeq \mathbb{C}^N \otimes_{\mathbb{C}} \mathcal{A}$. As already mentioned, in spite of the orthogonality of the basis elements, one has that $\|(a_1, \cdots, a_N)\|_{\mathcal{A}} =: \| \sum_{k=1}^n a_k^* a_k \| \neq \sum_{k=1}^n \|a_k^* a_k\|$. Parallel to the situation of the previous example, since the algebra $\mathcal{A}$ is unital, one finds that $End_{\mathcal{A}}(\mathcal{A}^N) \simeq End_{\mathcal{A}}^0(\mathcal{A}^N) \simeq \mathbb{M}_n(\mathcal{A})$. Here $\mathbb{M}_n(\mathcal{A})$ is the algebra of $n \times n$ matrices with entries in $\mathcal{A}$; it acts on the left on $\mathcal{A}^N$. The isometric isomorphism $End_{\mathcal{A}}^0(\mathcal{A}^N) \simeq \mathbb{M}_n(\mathcal{A})$ is now given by

$$End_{\mathcal{A}}^0(\mathcal{A}) \ni |(a_1, \cdots, a_N)\rangle \langle (b_1, \cdots, b_N)| \mapsto \begin{pmatrix} a_1 b_1^* & \cdots & a_1 b_N^* \\ \vdots & & \vdots \\ a_N b_1^* & \cdots & a_N b_N^* \end{pmatrix},$$

$$\forall \, a_k, b_k \in \mathcal{A}, \tag{A.38}$$

which is then extended by linearity.

Example A.3.3. The sections of a Hermitian complex vector bundle.
Let $\mathcal{A} = C(M)$ be the commutative C^*-algebra of complex-valued continuous functions on the locally compact Hausdorff space M. Here the norm is the sup norm as in (2.9). Given a complex vector bundle $E \to M$, the collection $\Gamma(E, M)$ of its continuous sections is a $C(M)$-module. This module is made into a Hilbert module if the bundle carries a Hermitian structure, namely a Hermitian scalar product $\langle \, , \, \rangle_{E_p} : E_p \times E_p \to \mathbb{C}$ on each fibre E_p, which varies continuously over M (as the space M is compact, this is always the case, any such structure is constructed by standard arguments with a partition of unit). The $C(M)$-valued Hermitian structure on $\Gamma(E, M)$ is then given by

$$\langle \eta_1, \eta_2 \rangle (p) = \langle \eta_1(p), \eta_2(p) \rangle_{E_p}, \quad \forall \, \eta_1, \eta_2 \in \Gamma(E, M), \, p \in M. \tag{A.39}$$

The module $\Gamma(E, M)$ is complete for the associated norm. It is also full since the linear span of $\{\langle \eta_1, \eta_2 \rangle, \, \eta_1, \eta_2 \in \Gamma(E, M)\}$ is dense in $C(M)$. Furthermore, as the module is projective of finite type, it follows from Proposition A.3.1 (see later) that

$$End_{C(M)}(\Gamma(E, M)) \simeq End_{C(M)}^0(\Gamma(E, M)) = \Gamma(EndE, M) \tag{A.40}$$

is the C^*-algebra of continuous sections of the endomorphism bundle $EndE \to M$ of E.

If M is only locally compact, one has to consider the algebra $C_0(M)$ of complex-valued continuous functions vanishing at infinity and the corresponding module $\Gamma_0(E, M)$ of continuous sections vanishing at infinity which again can be made into a full Hilbert module as before. But now one finds that $End_{C(M)}(\Gamma_0(E, M)) = \Gamma_b(EndE, M)$, the algebra of bounded sections, while $End^0_{C(M)}(\Gamma_0(E,M)) = \Gamma_0(EndE, M)$, the algebra of sections vanishing at infinity.

It is worth mentioning that not every Hilbert module over $C(M)$ arises in the manner described in the previous example. From the Serre-Swan theorem described in Sect. 4.2, one obtains only (and all) the projective modules of finite type. Now, there is a beautiful characterization of projective modules $\mathcal{E}$ over a C^*-algebra $\mathcal{A}$ in terms of the compact operators $End^0(\mathcal{E})$ [114, 106],

Proposition A.3.1. *Let $\mathcal{A}$ be a unital C^*-algebra.*

1. *Let $\mathcal{E}$ be a Hilbert module over $\mathcal{A}$ such that $\mathbb{I}_{\mathcal{E}} \in End^0(\mathcal{E})$ (so that $End(\mathcal{E}) = End^0(\mathcal{E})$). Then, the underlying right $\mathcal{A}$-module is projective of finite type.*
2. *Let $\mathcal{E}$ be a projective module of finite type over $\mathcal{A}$. Then, there exist $\mathcal{A}$-valued Hermitian structures on $\mathcal{E}$ for which $\mathcal{E}$ becomes a Hilbert module and one has that $\mathbb{I}_{\mathcal{E}} \in End^0(\mathcal{E})$. Furthermore, given any two $\mathcal{A}$-valued non degenerate Hermitian structures $\langle\ ,\ \rangle_1$ and $\langle\ ,\ \rangle_2$, on $\mathcal{E}$, there exists an invertible endomorphism T of $\mathcal{E}$ such that*

$$\langle\eta,\xi\rangle_2 = \langle T\eta, T\xi\rangle_1\ ,\quad \forall\ \eta,\xi \in \mathcal{E}\ . \tag{A.41}$$

Proof. To prove point 1., observe that by hypothesis there are two finite strings $\{\xi_k\}$ and $\{\zeta_k\}$ of elements of $\mathcal{E}$ such that

$$\mathbb{I}_{\mathcal{E}} = \sum_k |\xi_k\rangle\langle\zeta_k|\ . \tag{A.42}$$

Then, for any $\eta \in \mathcal{E}$, one has that

$$\eta = \mathbb{I}_{\mathcal{E}}\eta = \sum_k |\xi_k\rangle\langle\zeta_k|\eta = \sum_k \xi_k \langle\zeta_k,\eta\rangle_{\mathcal{A}}\ , \tag{A.43}$$

and hence $\mathcal{E}$ is finitely generated by the string $\{\xi_k\}$. If N is the length of the strings $\{\xi_k\}$ and $\{\zeta_k\}$, one can embed $\mathcal{E}$ as a direct summand of $\mathcal{A}^N$, proving that $\mathcal{E}$ is projective. The embedding and the surjection maps are defined, respectively, by

$$\lambda : \mathcal{E} \to \mathcal{A}^N\ ,\quad \lambda(\eta) = (\langle\zeta_1,\eta\rangle_{\mathcal{A}},\cdots,\langle\zeta_N,\eta\rangle_{\mathcal{A}})\ ,$$
$$\rho : \mathcal{A}^N \to \mathcal{E}\ ,\quad \rho((a_1,\cdots,a_N)) = \sum_k \xi_k a_k\ . \tag{A.44}$$

Then, for any $\eta \in \mathcal{E}$, $\rho\circ\lambda(\eta) = \rho((\langle\zeta_1,\eta\rangle_{\mathcal{A}},\cdots,\langle\zeta_N,\eta\rangle_{\mathcal{A}})) = \sum_k \xi_k \langle\zeta_k,\eta\rangle_{\mathcal{A}} = \sum_k |\xi_k\rangle\langle\zeta_k|(\eta) = \mathbb{I}_{\mathcal{E}}(\eta)$, so that $\rho\circ\lambda = \mathbb{I}_{\mathcal{E}}$ as required. The projector $p = \lambda\circ\rho$ identifies $\mathcal{E}$ as $p\mathcal{A}^N$.

To prove point 2., observe that, as the module $\mathcal{E}$ is a direct summand of the free module $\mathcal{A}^N$ for some N, the restriction the Hermitian structure (A.36) on the latter to the submodule $\mathcal{E}$ gives it a Hilbert module structure. Furthermore, if $\rho : \mathcal{A}^N \to \mathcal{E}$ is the surjection associated with $\mathcal{E}$, the image $\epsilon_k = \rho(e_k), k = 1, \ldots N$, of the free basis $\{e_k\}$ of $\mathcal{A}^N$ described in Example A.3.2 is a (not free) basis of $\mathcal{E}$. Then the identity $\mathbb{I}_{\mathcal{E}}$ can be written as

$$\mathbb{I}_{\mathcal{E}} = \sum_k |\epsilon_k\rangle \langle \epsilon_k| \, , \qquad (A.45)$$

and is an element of $End^0_{\mathcal{A}}(\mathcal{E})$.

Finally, from Sect. 4.3, given two Hermitian structures $\langle \, , \, \rangle_i$, $i = 1, 2$ on $\mathcal{E}$, there exist maps $Q_i : \mathcal{E} \to \mathcal{E}' =: Hom_{\mathcal{A}}(\mathcal{E}, \mathcal{A})$, defined by

$$Q_i(\xi)(\eta) =: \langle \xi, \eta \rangle_i \, , \ i = 1, 2 \, , \ \forall \, \xi, \eta \in \mathcal{E} \, . \qquad (A.46)$$

One has that $Q_i(\xi a) = a^* Q_i(\xi)$, for any $a \in \mathcal{A}$ (remember the right $\mathcal{A}$-module structure given to $\mathcal{E}'$ in (4.21)). The non degeneracy of the Hermitian structures is equivalent to the invertibility of the two maps Q_i. Furthermore, since the Hermitian structures are both positive (from (A.20)), the invertible endomorphism $Q_1^{-1} \circ Q_2$ of $\mathcal{E}$ admits a square root[2], $T = \sqrt{Q_1^{-1} \circ Q_2}$. Then for any $\xi, \eta \in \mathcal{E}$, it follows that

$$\begin{aligned} \langle T\xi, T\eta \rangle_1 &= \langle TT\xi, \eta \rangle_1 = Q_1(T^2\xi)(\eta) \\ &= Q_1(Q_1^{-1} \circ Q_2\xi)(\eta) = Q_2(\xi)(\eta) \\ &= \langle T\xi, T\eta \rangle_2 \, . \end{aligned} \qquad (A.47)$$

---

[2] In fact, one needs a technical requirement, namely that C^*-algebra $\mathcal{A}$ be stable under holomorphic functional calculus. We recall that this means that for any $a \in \mathcal{A}$ and any function f which is holomorphic in a neighborhood of the spectrum of a, one has that $f(a) \in \mathcal{A}$.

A.4 Strong Morita Equivalence

In this Appendix, we describe the notion of strong Morita equivalence [113, 114] between two C^*-algebras. This really boils down to an equivalence between the corresponding representation theories. We refer to Appendix A.3 for the fundamentals of Hilbert modules over a C^*-algebra.

Definition A.4.1. *Let* A *and* B *be two* C^*-algebras. *We say that they are strongly Morita equivalent if there exists a* B-A *equivalence Hilbert bimodule* $\mathcal{E}$, *namely a module* $\mathcal{E}$ *which is at the same time a right Hilbert module over* A *with* A-valued Hermitian structure $\langle \, , \, \rangle_A$, *as well as being a left Hilbert module over* B *with* B-valued Hermitian structure $\langle \, , \, \rangle_B$ *such that*

1. *The module* $\mathcal{E}$ *is full both as a right and as a left Hilbert module;*
2. *The Hermitian structures are compatible,*

$$\langle \eta, \xi \rangle_B \, \zeta = \eta \, \langle \xi, \zeta \rangle_A \, , \quad \forall \, \eta, \xi, \zeta \in \mathcal{E} \, ; \tag{A.48}$$

3. *The left representation of* B *on* $\mathcal{E}$ *is a continuous* *-representation by operators which are bounded for $\langle \, , \, \rangle_A$, i.e. $\langle b\eta, b\eta \rangle_A \leq ||b||^2 \, \langle \eta, \eta \rangle_A$. *The right representation of* A *on* $\mathcal{E}$ *is a continuous* *-representation by operators which are bounded for $\langle \, , \, \rangle_B$, namely $\langle \eta a, \eta a \rangle_B \leq ||a||^2 \, \langle \eta, \eta \rangle_B$.

Example A.4.1. For any full Hilbert module $\mathcal{E}$ over the C^*-algebra A, the latter is strongly Morita equivalent to the C^*-algebra $End^0_A(\mathcal{E})$ of compact endomorphisms of $\mathcal{E}$. If $\mathcal{E}$ is projective of finite type, so that by Proposition A.3.1 $End^0_A(\mathcal{E}) = End_A(\mathcal{E})$, the algebra A is strongly Morita equivalent to the whole of $End_A(\mathcal{E})$.

Consider then a full *right* Hilbert module $\mathcal{E}$ on the algebra A with A-valued Hermitian structure $\langle \, , \, \rangle_A$. Now, $\mathcal{E}$ is a *left* module over the C^*-algebra $End^0_A(\mathcal{E})$. A left Hilbert module structure is constructed by inverting definition (A.27) so as to produce an $End^0_A(\mathcal{E})$-valued Hermitian structure on $\mathcal{E}$,

$$\langle \eta_1, \eta_2 \rangle_{End^0_A(\mathcal{E})} =: |\eta_1\rangle \langle \eta_2| \, , \quad \forall \, \eta_1, \eta_2 \in \mathcal{E} \, . \tag{A.49}$$

It is straightforward to check that the previous structure satisfies all the properties of a left structure including conjugate linearity in the second variable. From the definition of compact endomorphisms, the module $\mathcal{E}$ is also full as a module over $End^0_A(\mathcal{E})$ so that requirement 1. in the Definition A.4.1 is satisfied. Furthermore, from definition A.27 one has that for any $\eta_1, \eta_2, \xi \in \mathcal{E}$,

$$\langle \eta_1, \eta_2 \rangle_{End^0_A(\mathcal{E})} \, \xi =: |\eta_1\rangle \langle \eta_2| (\xi) = \eta_1 \, \langle \eta_2, \xi \rangle_A \, , \tag{A.50}$$

so that also requirement 2. is met. Finally, the left action of $End^0_A(\mathcal{E})$ on $\mathcal{E}$ as an A-module is by bounded operators. And, finally, for any $a \in A$, $\eta, \xi \in \mathcal{E}$, one has that

$$\Big\langle \langle \eta a, \eta a \rangle_{End_{\mathcal{A}}^0(\mathcal{E})} \, \xi, \xi \Big\rangle_{\mathcal{A}} \;=\; \langle (\eta a) \, \langle \eta a, \xi \rangle_{\mathcal{A}} \, , \xi \rangle_{\mathcal{A}}$$

$$=\; \langle \eta a a^* \, \langle \eta, \xi \rangle_{\mathcal{A}} \, , \xi \rangle_{\mathcal{A}}$$

$$=\; \langle \eta, \xi \rangle_{\mathcal{A}}^* \, a a^* \, \langle \eta, \xi \rangle_{\mathcal{A}}$$

$$\leq\; ||a||^2 \, \langle \eta, \xi \rangle_{\mathcal{A}}^* \, \langle \eta, \xi \rangle_{\mathcal{A}}$$

$$\leq\; ||a||^2 \, \langle \eta \, \langle \eta, \xi \rangle_{\mathcal{A}} \, , \xi \rangle_{\mathcal{A}}$$

$$\leq\; ||a||^2 \, \Big\langle \langle \eta, \eta \rangle_{End_{\mathcal{A}}^0(\mathcal{E})} \, \xi, \xi \Big\rangle_{\mathcal{A}} \, , \qquad \text{(A.51)}$$

from which we find

$$\langle \eta a, \eta a \rangle_{End_{\mathcal{A}}^0(\mathcal{E})} \leq ||a||^2 \, \langle \eta, \eta \rangle_{End_{\mathcal{A}}^0(\mathcal{E})} \, , \qquad \text{(A.52)}$$

which is the last requirement of Definition A.4.1.

Given any $\mathcal{B}$-$\mathcal{A}$ equivalence Hilbert bimodule $\mathcal{E}$ one can exchange the rôle of $\mathcal{A}$ and $\mathcal{B}$ by constructing the associated *complex conjugate*[3] $\mathcal{A}$-$\mathcal{B}$ equivalence Hilbert bimodule $\widetilde{\mathcal{E}}$ with a *right* action of $\mathcal{B}$ and a *left* action of $\mathcal{A}$. As an additive group $\widetilde{\mathcal{E}}$ is identified with $\mathcal{E}$ and any element of it will be denoted by $\widetilde{\eta}$, with $\eta \in \mathcal{E}$. Then one gives a conjugate action of $\mathcal{A}, \mathcal{B}$ (and complex numbers) with corresponding Hermitian structures. The left action by $\mathcal{A}$ and the right action by $\mathcal{B}$ are defined by

$$a \cdot \widetilde{\eta} =: \widetilde{\eta a^*} \, , \quad \forall \, a \in \mathcal{A} \, , \widetilde{\eta} \in \widetilde{\mathcal{E}} \, , \qquad \text{(A.53)}$$

$$\widetilde{\eta} \cdot b =: \widetilde{b^* \eta} \, , \quad \forall \, b \in \mathcal{B} \, , \widetilde{\eta} \in \widetilde{\mathcal{E}} \, , \qquad \text{(A.54)}$$

and are readily seen to satisfy the appropriate properties. As for the Hermitian structures, they are given by

$$\langle \widetilde{\eta}_1, \widetilde{\eta}_2 \rangle_{\mathcal{A}} =: \langle \eta_1, \eta_2 \rangle_{\mathcal{A}} \, , \qquad \text{(A.55)}$$

$$\langle \widetilde{\eta}_1, \widetilde{\eta}_2 \rangle_{\mathcal{B}} =: \langle \eta_1, \eta_2 \rangle_{\mathcal{B}} \, , \quad \forall \, \widetilde{\eta}_1, \widetilde{\eta}_2 \in \mathcal{E} \, . \qquad \text{(A.56)}$$

Again one readily checks that the appropriate properties, notably conjugate linearity in the second and first variable respectively, are satisfied as well as all the other requirements for an $\mathcal{A}$-$\mathcal{B}$ equivalence Hilbert bimodule.

As already mentioned, two strongly Morita equivalent C^*-algebras have equivalent representation theories. We sketch this fact in the what follows while referring to [113, 114] for more details.

Suppose then that we are given two strongly Morita equivalent C^*-algebras $\mathcal{A}$ and $\mathcal{B}$ with $\mathcal{B}$-$\mathcal{A}$ equivalence bimodule $\mathcal{E}$. Let $(\mathcal{H}, \pi_{\mathcal{A}})$ be a representation of $\mathcal{A}$ on the Hilbert space $\mathcal{H}$. The algebra $\mathcal{A}$ acts with bounded operators on the left on $\mathcal{H}$ via π. This action can be used to construct another Hilbert space

---

[3] Not to be confused with the dual module introduced in eq. (4.21).

$$\mathcal{H}' =: \mathcal{E} \otimes_{\mathcal{A}} \mathcal{H} , \quad \eta a \otimes_{\mathcal{A}} \psi - \eta \otimes_{\mathcal{A}} \pi_{\mathcal{A}}(a)\psi = 0 , \quad \forall\, a \in \mathcal{A},\ \eta \in \mathcal{E},\ \psi \in \mathcal{H} ,$$
(A.57)

with scalar product

$$(\eta_1 \otimes_{\mathcal{A}} \psi_1, \eta_2 \otimes_{\mathcal{A}} \psi_2) =: (\psi_1, \langle \eta_1, \eta_2 \rangle_{\mathcal{A}}\, \psi_2)_{\mathcal{H}} , \quad \forall\, \eta_1, \eta_2 \in \mathcal{E},\ \psi_1, \psi_2 \in \mathcal{H} .$$
(A.58)

A representation $(\mathcal{H}', \pi_{\mathcal{B}})$ of the algebra $\mathcal{B}$ is constructed by

$$\pi_{\mathcal{B}}(b)(\eta \otimes_{\mathcal{A}} \psi) =: (b\eta) \otimes_{\mathcal{A}} \psi , \quad \forall\, b \in \mathcal{A},\ \eta \otimes_{\mathcal{A}} \psi \in \mathcal{H}' .$$
(A.59)

This representation is unitary equivalent to the representation $(\mathcal{H}, \pi_{\mathcal{A}})$. If one starts with a representation of $\mathcal{B}$, by using the conjugate $\mathcal{A}$-$\mathcal{B}$ equivalence bimodule $\widetilde{\mathcal{E}}$ one constructs an equivalent representation of $\mathcal{A}$. Therefore, there is an equivalence between the category of representations of the algebra $\mathcal{A}$ and the category of representations of the algebra $\mathcal{B}$

As a consequence, strong Morita equivalent C^*-algebras $\mathcal{A}$ and $\mathcal{B}$ have the same space of classes of (unitary equivalent) irreducible representations. Furthermore, there also exists an isomorphism between the lattice of two-sided ideals of $\mathcal{A}$ and $\mathcal{B}$ and a homeomorphism between the spaces of primitive ideals of $\mathcal{A}$ and $\mathcal{B}$.

In particular, if a C^*-algebra $\mathcal{A}$ is strongly Morita equivalent to some commutative C^*-algebra, from the results of Sect. 2.2, the latter is unique and is the C^*-algebra of continuous functions vanishing at infinity on the space M of irreducible representations of $\mathcal{A}$.

For any integer n, the algebra $\mathbb{M}_n(\mathbb{C}) \otimes C_0(M) \simeq \mathbb{M}_n(C_0(M))$ is strongly Morita equivalent to the algebra $C_0(M)$. In particular, the algebras $\mathbb{M}_n(\mathbb{C})$ and $\mathbb{C}$ are strongly Morita equivalent.

It is worth mentioning that if $\mathcal{A}$ and $\mathcal{B}$ are two separable C^*-algebras and $\mathcal{K}$ is the C^*-algebra of compact operators on an infinite dimensional separable Hilbert space, then one proves [22] that the algebras $\mathcal{A}$ and $\mathcal{B}$ are strongly Morita equivalent if and only if $\mathcal{A} \otimes \mathcal{K}$ is isomorphic to $\mathcal{B} \otimes \mathcal{K}$.

In [112], $\mathcal{B}$-$\mathcal{A}$ equivalence Hilbert bimodules have been used to derive a very nice formulation of spinor fields.

A.5 Partially Ordered Sets

Here we gather together some facts about partially ordered set. These are mainly taken from [122].

Definition A.5.1. *A* partially ordered set *(or* poset *for short) P is a set endowed with a binary relation $\preceq$ which satisfies the following axioms:*

P_1 $x \preceq x$,   *for all $x \in P$;* *(*reflexivity*)*
P_2 $x \preceq y$ *and* $y \preceq x$ $\Rightarrow$ $x = y$; *(*antisymmetry*)*
P_3 $x \preceq y$ *and* $y \preceq z$ $\Rightarrow$ $x \preceq z$. *(*transitivity*)*

The relation $\preceq$ is called a *partial order* and the set P will be said to be partially ordered. The relation $x \preceq y$ is also read 'x precedes y'. The obvious notation $x \prec y$ will mean $x \preceq y$ and $x \neq y$; $x \succeq y$ will mean $y \preceq x$ and $x \succ y$ will mean $y \prec x$. Two elements x and y of P are said to be *comparable* if $x \preceq y$ or $y \preceq x$; otherwise they are *incomparable* (or *not comparable*). A subset Q of P is called a *subposet* of P if it is endowed with the induced order, namely for any $x, y \in Q$ one has $x \preceq_Q y$ in Q if and only if $x \preceq_P y$ in P.

An element $x \in P$ is called *maximal* if there is no other $y \in P$ such that $x \prec y$. An element $x \in P$ is called *minimal* if there is no other $y \in P$ such that $y \prec x$. Notice that P may admit more that one maximal and/or minimal point. One says that P admits a $\hat{0}$ if there exists an element $\hat{0} \in P$ such that $\hat{0} \preceq x$ for all $x \in P$. Similarly, P admits a $\hat{1}$ if there exists an element $\hat{1} \in P$ such that $x \preceq \hat{1}$ for all $x \in P$.

Example A.5.1. Any collection of sets can be partially ordered by inclusion. In particular, throughout the paper we have considered the collection of all primitive ideals of a C^*-algebra at length.

Example A.5.2. As mentioned in the previous Appendix, the set of all possible topologies on the same space S is a partially ordered set. If τ_1 and τ_2 are two topologies on the space S, one puts $\tau_1 \preceq \tau_2$ if and only if τ_1 is coarser than τ_2. The corresponding poset has a $\hat{0}$, the coarsest topology, in which only $\emptyset$ and S are open, and a $\hat{1}$, the finest topology, in which all subsets of S are open.

Two posets P and Q are *isomorphic* if there exists an *order preserving bijection* $\phi : P \rightarrow Q$, that is $x \preceq y$ in P if and only if $\phi(x) \preceq \phi(y)$ in Q, whose inverse is also order preserving.

For any relation $x \preceq y$ in P, we get a (closed) interval defined by $[x, y] = \{z \in P \mid x \preceq z \preceq y\}$. The poset P is called *locally finite* if every interval of P is finite (it consists of a finite number of elements).

If x and y are in P, we say that y covers x if $x \prec y$ and no element $z \in P$ satisfies $x \prec z \prec y$. A locally finite poset is completely determined by its cover relations.

The *Hasse diagram* of a (finite) poset P is a graph whose vertices are the elements of P drawn in such a way that if $x \prec y$ then y is 'above' x; furthermore, the *links* are the cover relations, namely, if y covers x then a link is drawn between x and y. One does not draw links which would be implied by transitivity. In Chap. 3 a few Hasse diagrams were given.

A *chain* is a poset in which any two elements are comparable. A subset C of a poset P is called a chain (of P) if C is a chain when regarded as a subposet of P. The *length* $\ell(C)$ of a finite chain is defined as $\ell(C) = |C| - 1$, with $|C|$ the number of elements in C. The *length* (or *rank*) of a finite poset P is defined as $\ell(P) =: \max \{\ell(C) \mid$ is a chain of $P\}$. If every maximal chain of P has the same length n, one says that P is *graded of rank n*. In this case there is a unique *rank function* $\rho : P \to \{0, 1, \ldots, n\}$ such that $\rho(x) = 0$ if x is a minimal element and $\rho(y) = \rho(x) + 1$, if y covers x. The point $x \in P$ is said to be of *rank i* if $\rho(x) = i$.

If P and Q are posets, their *cartesian product* is the poset $P \times Q$ on the set $\{(x, y) \mid x \in P, y \in Q\}$ such that $(x, y) \preceq (x', y')$ in $P \times Q$ if $x \preceq x'$ in P and $y \preceq y'$ in Q. To draw the Hasse diagram of $P \times Q$, one draws the diagram of P, replaces each element x of P by a copy Q_x of Q and connects corresponding elements of Q_x and Q_y (by identifying $Q_x \simeq Q_y$) if x and y are connected in the diagram of P.

Finally we mention that the *dual* of a poset P is the poset P^* on the same set as P, but such that $x \preceq y$ in P^* if and only if $y \preceq x$ in P. If P and P^* are isomorphic, then P is called *self-dual*.

If x and y belong to a poset P, an *upper bound* of x and y is an element $z \in P$ for which $x \preceq z$ and $y \preceq z$. A *least upper bound* of x and y is an upper bound z of x and y such that any other upper bound w of x and y satisfies $z \preceq w$. If a least upper bound of x and y exists, then it is unique and it is denoted $x \vee y$, '*x join y*'. Dually one can define the greatest lower bound $x \wedge y$, '*x meet y*', when it exists. A *lattice* is a poset L for which every pair of elements has a join and a meet. In a lattice the operations $\vee$ and $\wedge$ satisfy the following properties

1. they are associative, commutative and idempotent
   $(x \vee x = x \wedge x = x)$;
2. $x \wedge (x \vee y) = x = x \vee (x \wedge y)$ (absorbation laws);
3. $x \wedge y = x \Leftrightarrow x \vee y \Leftrightarrow x \preceq y$.

All finite lattices have the element $\hat{0}$ and the element $\hat{1}$.

A.6 Pseudodifferential Operators

We shall give a very sketchy overview of some aspects of the theory of pseudo differential operators while referring to [96, 126] for details.

Suppose we are given a rank k vector bundle $E \to M$ with M a compact manifold of dimension n. We shall denote by $\Gamma(E)$ the $C^\infty(M)$-module of corresponding smooth sections.

A *differential operator of rank m* is a linear operator

$$P : \Gamma(M) \longrightarrow \Gamma(M) , \tag{A.60}$$

which, in local coordinates $x = (x_1, \cdots, x_n)$ of M, is written as

$$P = \sum_{|\alpha| \le m} A_\alpha(x)(-i)^{|\alpha|}\frac{\partial^{|\alpha|}}{\partial x^\alpha} , \quad \frac{\partial^{|\alpha|}}{\partial x^\alpha} = \frac{\partial^{\alpha_1}}{\partial x_1^{\alpha_1}} \circ \cdots \circ \frac{\partial^{\alpha_n}}{\partial x_1^{\alpha_n}} . \tag{A.61}$$

Here $\alpha = (\alpha_1, \cdots, \alpha_n), 0 \le \alpha_j \le n$, is a multi-index of cardinality $|\alpha| = \sum_{j=1}^n \alpha_j$. Each A_α is a $k \times k$ matrix of smooth functions on M and $A_\alpha \ne 0$ for some α with $|\alpha| = m$.

Consider now an element ξ of the cotangent space T_x^*M, $\xi = \sum_j \xi_j dx_j$. The *complete symbol* of P is defined by the following polynomial function in the components ξ_j.

$$p^P(x,\xi) = \sum_{j=0}^m p_{m-j}^P(x,\xi) , \quad p_{m-j}^P(x,\xi) = \sum_{|\alpha| \le (m-j)} A_\alpha(x)\xi^\alpha , \tag{A.62}$$

and the leading term is called the *principal symbol*

$$\sigma^P(x,\xi) = p_m^P(x,\xi) = \sum_{|\alpha|=m} A_\alpha(x)\xi^\alpha , \tag{A.63}$$

here $\xi^\alpha = \xi_1^{\alpha_1} \cdots \xi_n^{\alpha_n}$. Hence, for each cotangent vector $\xi \in T_x^*M$, the principal symbol gives a map

$$\sigma^P(\xi) : E_x \longrightarrow E_x , \tag{A.64}$$

where E_x is the fibre of E over x. If $\tau : T^*M \to M$ is the cotangent bundle of M and τ^*E the pullback of the bundle E to T^*M, then, the principal symbol σ^P determines in an invariant manner a (fibre preserving) bundle homomorphism of τ^*E, namely an element of $\Gamma(\tau^*EndE \to T^*M)$.

The differential operator P is called *elliptic* if its principal symbol $\sigma^P(\xi) : E_x \to E_x$ is invertible for any non zero cotangent vector $\xi \in T^*M$. If M is a Riemannian manifold with metric $g = (g^{\mu\nu})$, since $\sigma^P(\xi)$ is polynomial in ξ, being elliptic is equivalent to the fact that the linear transformation $\sigma^P(\xi) : E_x \to E_x$ is invertible on the cosphere bundle

$$S^*M = \{(x,\xi) \in T^*M \mid g^{\mu\nu}\xi_\mu\xi_\nu = 1\} . \tag{A.65}$$

Example A.6.1. The Laplace-Beltrami operator $\Delta : C^\infty(M) \to C^\infty(M)$ of a Riemannian metric $g = (g_{\mu\nu})$ on M, in local coordinates is written as

$$\Delta f = -\sum_{\mu\nu} g^{\mu\nu} \frac{\partial^2 f}{\partial x^\mu \partial x^\nu} + \text{ lower order terms .} \tag{A.66}$$

As for its principal symbol we have,

$$\sigma^\Delta(\xi) = \sum_{\mu\nu} g^{\mu\nu} \xi_\mu \xi_\nu = ||\xi||^2 , \tag{A.67}$$

which is clearly invertible for any non zero cotangent vector ξ. Therefore, the Laplace-Beltrami operator is an elliptic second order differential operator.

Example A.6.2. Suppose now that M is a Riemannian spin manifold as in Sect. 6.5. The corresponding Dirac operator can be written locally as,

$$D = \gamma(dx^\mu)\partial_\mu + \text{ lower order terms ,} \tag{A.68}$$

were γ is the algebra morphism defined in (6.48). Then, its principal symbol is just the 'Clifford multiplication' by ξ,

$$\sigma^D(\xi) = \gamma(\xi) . \tag{A.69}$$

By using (6.49) one has $\gamma(\xi)^2 = -||\xi||^2 Id$, and the symbol is certainly invertible for $\xi \neq 0$. Therefore, the Dirac operator is an elliptic first order differential operator.

By using its symbol, the action of the operator P on a local section u of the bundle E can be written as a Fourier integral,

$$(Pu)(x) = \frac{1}{(2\pi)^{n/2}} \int e^{i\langle \xi, x \rangle} p(x, \xi) \hat{u}(\xi) d\xi ,$$

$$\hat{u}(\xi) = \frac{1}{(2\pi)^{n/2}} \int e^{-i\langle \xi, x \rangle} u(x) dx , \tag{A.70}$$

with $\langle \xi, x \rangle = \sum_{j=1}^n \xi_j x_j$.

One uses the formula (A.70) to define *pseudodifferential operators*, taking $p(x, \xi)$ to belong to a more general class of symbols. The problem is to control the growth of powers in k. We shall suppose, for simplicity, that we have a trivial vector bundle over $\mathbb{R}^n$ of rank k.

With $m \in \mathbb{R}$, one defines the symbol class Sym^m to consist of matrix-valued smooth functions $p(x, \xi)$ on $\mathbb{R}^n \times \mathbb{R}^n$, with the property that, for any x-compact $K \subset \mathbb{R}^n$ and any multi-indices α, β, there exists a constant $C_{K\alpha\beta}$ such that

$$|D_x^\beta D_\xi^\alpha p(x, \xi)| \leq C_{K\alpha\beta}(1 + |\xi|)^{m-|\alpha|}, \tag{A.71}$$

with $D_x^\beta = (-i)^{|\beta|}\partial^{|\beta|}/\partial x^\beta$ and $D_\xi^\alpha = (-i)^{|\alpha|}\partial^{|\alpha|}/\partial\xi^\alpha$. Furthermore, the function $p(x,\xi)$ has an 'asymptotic expansion' given by

$$p(x,\xi) \sim \sum_{j=0}^{\infty} p_{m-j}(x,\xi) \ . \tag{A.72}$$

where p_{m-j} are matrices of smooth functions on $\mathbb{R}^n \times \mathbb{R}^n$, homogeneous in ξ of degree $(m-j)$,

$$p_{m-j}(x,\lambda\xi) = \lambda^{m-j} p_{m-j}(x,\xi) \ , \quad |\xi| \geq 1, \ \lambda \geq 1 \ . \tag{A.73}$$

The asymptotic condition (A.72) means that for any integer N, the difference

$$p(x,\xi) - \sum_{j=0}^{N} p_{m-j}(x,\xi) = F^N(x,\xi) \tag{A.74}$$

satisfies a regularity condition similar to (A.71): for any x-compact $K \in \mathbb{R}^n$ and any multi-indices α, β there exists a constant $C_{K\alpha\beta}$ such that

$$|D_x^\beta D_\xi^\alpha F^N(x,\xi)| \leq C_{K\alpha\beta}(1 + |\xi|)^{m-(N+1)-|\alpha|} \ . \tag{A.75}$$

Thus, $F^N \in Sym^{m-N-1}$ for any integer N.

As we said before, any symbol $p(x,\xi) \in Sym^m$ defines a pseudodifferential operator P of order m by formula (A.70) where now u is a section of the rank k trivial bundle over $\mathbb{R}^n$ and can therefore be identified with a $\mathbb{C}^k$-valued smooth function on $\mathbb{R}^n$. The space of all such operators is denoted by ΨDO_m. Let $P \in \Psi DO_m$ with symbol $p \in Sym^m$. Then, the *principal symbol* of P is the residue class $\sigma^P = [p] \in Sym^m/Sym^{m-1}$. One can prove that the principal symbol transforms under diffeomorphisms as a matrix-valued function on the cotangent bundle of $\mathbb{R}^n$.

The class $Sym^{-\infty}$ is defined by $\bigcap_m Sym^m$ and the corresponding operators are called *smoothing operators*, the space of all such operators being denoted by $\Psi DO_{-\infty}$. A smoothing operator S has an integral representation with a smooth kernel which means that its action on a section u can be written as

$$(Pu)(x) = \int K(x,y)u(y)dy \ , \tag{A.76}$$

where $K(x,y)$ is a smooth function on $\mathbb{R}^n \times \mathbb{R}^n$ (with compact support). One is really interested in equivalence classes of pseudodifferential operators, where two operators P and P' are declared equivalent if $P - P'$ is a smoothing operator.

Given $P \in \Psi DO_m$ and $Q \in \Psi DO_\mu$ with symbols $p(x,\xi)$ and $q(x,\xi)$ respectively, the composition $R = P \circ Q \in \Psi DO_{m+\mu}$ has symbol with asymptotic expansion

$$r(x,\xi) \sim \sum_\alpha \frac{i^{|\alpha|}}{\alpha!} D_\xi^\alpha p(x,\xi) D_x^\alpha q(x,\xi) \ . \tag{A.77}$$

In particular, the leading term $|\alpha| = 0$ in the previous expression shows that the principal symbol of the composition is the product of the principal symbols of the factors

$$\sigma^R(x,\xi) = \sigma^P(x,\xi)\sigma^Q(x,\xi) \ . \tag{A.78}$$

Given $P \in \Psi DO_m$, its formal adjoint P^* is defined by

$$(Pu, v)_{L^2} = (u, P^*)_{L^2}, \tag{A.79}$$

for all sections u, v with compact support. Then, $P^* \in \Psi DO_m$ and, if P has symbol $p(x,\xi)$, the operator P^* has symbol $p^*(x,\xi)$ with asymptotic expansion

$$p^*(x,\xi) \sim \sum_\alpha \frac{i^{|\alpha|}}{\alpha!} D_\xi^\alpha D_x^\alpha (p(x,\xi))^* \ , \tag{A.80}$$

with the operation $*$ on the right-hand side denoting Hermitian matrix conjugation $(p(x,\xi))^* = \overline{p(x,\xi)}\,^t$, ^{t} being matrix transposition. Again, by taking the leading term $|\alpha| = 0$, we see that the principal symbol σ^{P^*} of P^* is just the Hermitian conjugate $(\sigma^P)^*$ of the principal symbol of P. As a consequence, the principal symbol of a positive pseudodifferential operator $R = P^*P$ is nonnegative.

An operator $P \in \Psi DO_m$ with symbol $p(x,\xi)$ is said to be *elliptic* if its principal symbol $\sigma^P \in Sym^m/Sym^{m-1}$ has a representative which, as a matrix-valued function on $T^*\mathbb{R}^n$ is pointwise invertible outside the zero section $\xi = 0$ in $T^*\mathbb{R}^n$. An elliptic (pseudo-)differential operator $P \in \Psi DO_m$ admits an inverse modulo smoothing operators. This means that there exists a pseudodifferential operator $Q \in \Psi DO_{-m}$ such that

$$PQ - \mathbb{I} = S_1 \ ,$$
$$QP - \mathbb{I} = S_2 \ , \tag{A.81}$$

with S_1 and S_2 smoothing operators. The operator Q is called a *paramatrix* for P.

The general situation of pseudodifferential operators acting on sections of a nontrivial vector bundle $E \to M$, with M compact, is worked out with suitable partitions of unity. An operator P acting on $\Gamma(E \to M)$ is a pseudodifferential operator of order m, if and only if the operator $u \mapsto \phi P(\psi u)$ is a pseudodifferential operator of order m for any $\phi, \psi \in C^\infty(M)$ which are supported in trivializing charts for E. The operator P is then recovered from its components via a partition of unity. Although the symbol of the operator P will depend on the charts, just as for ordinary differential operators, its principal symbol σ^P has an invariant meaning as a mapping from T^*M into endomorphisms of $E \to M$. Thus, ellipticity has an invariant meaning and an operator P is called elliptic if its principal symbol σ^P is pointwise invertible off the zero section of T^*M. Again, if M is a Riemannian manifold with metric $g = (g^{\mu\nu})$, since $\sigma^P(\xi)$ is homogeneous in ξ, being elliptic means that the linear transformation $\sigma^P(\xi) : E_x \to E_x$ is invertible on the cosphere bundle $S^*M \subset T^*M$.

Example A.6.3. Consider the one dimensional Hamiltonian given, in 'momentum space' by

$$H(\xi, x) = \xi^2 + V(x) \,, \tag{A.82}$$

with $V(x) \in C^\infty(\mathbb{R})$. It is clearly a differential operator of order 2. The following are associated pseudodifferential operators of order $-2, 1, -1$ respectively [42],

$$(\xi^2 + V)^{-1} = \xi^{-2} - V\xi^{-4} + 2V^{(1)}\xi^{-5} + \dots \,,$$

$$(\xi^2 + V)^{1/2} = \xi + \frac{V}{2}\xi^{-1} - \frac{V^{(1)}}{4}\xi^{-2} + \dots \,,$$

$$(\xi^2 + V)^{-1/2} = \xi^{-1} - \frac{V}{2}\xi^{-3} + \frac{3V^{(1)}}{4}\xi^{-4} + \dots \,, \tag{A.83}$$

where $V(k)$ is the k-th derivative of V with respect to its argument.
In particular, for the one dimensional harmonic oscillator $V(x) = x^2$. The pseudodifferential operators in (A.83) become,

$$(\xi^2 + x^2)^{-1} = \xi^{-2} - x^2\xi^{-4} + 4x\xi^{-5} + \dots \,,$$

$$(\xi^2 + x^2)^{1/2} = \xi + \frac{x^2}{2}\xi^{-1} - \frac{x}{2}\xi^{-2} + \dots \,.$$

$$(\xi^2 + x^2)^{-1/2} = \xi^{-1} - \frac{x^2}{2}\xi^{-3} + \frac{3x}{2}\xi^{-4} + \dots \,. \tag{A.84}$$

References

1. R. Abraham, J.E. Marsden, T. Ratiu, *Manifolds, Tensor Analysis and Applications* (Addison-Wesley, 1983).
2. T. Ackermann, *A Note on the Wodzicki Residue*, J. Geom. Phys. **20** (1996) 404-406.
3. S.L. Adler, *Einstein Gravity as a Symmetry-breaking Effect in Quantum Field Theory*, Rev. Mod. Phys. **54** (1982) 729-766.
4. P.S. Aleksandrov, *Combinatorial Topology*, Vol. I (Greylock, 1956).
5. M.F. Atiyah, R. Bott, A. Shapiro, *Clifford Modules*, Topology, **3** (1964), Supp. 1, 3-33.
6. A.P. Balachandran, G. Bimonte, E. Ercolessi, G. Landi, F. Lizzi, G. Sparano, P. Teotonio-Sobrinho, *Noncommutative Lattices as Finite Approximations*, J. Geom. Phys. **18** (1996) 163-194; (hep-th/9510217).
   A.P. Balachandran, G. Bimonte, E. Ercolessi, G. Landi, F. Lizzi, G. Sparano, P. Teotonio-Sobrinho, *Finite Quantum Physics and Noncommutative Geometry*, Nucl. Phys. B (Proc. Suppl.) **37C** (1995), 20-45; (hep-th/9403067).
7. A.P. Balachandran, G. Bimonte, G. Landi, F. Lizzi, P. Teotonio-Sobrinho, *Lattice Gauge Fields and Noncommutative Geometry*, J. Geom. Phys., to appear; (hep-lat/9604012).
8. H. Behncke, H. Leptin, C^*-*algebras with a Two-Point Dual*, J. Functional Analysis **10** (1972) 330-335;
   H. Behncke, H. Leptin, C^*-*algebras with Finite Duals*, J. Functional Analysis **14** (1973) 253-268;
   H. Behncke, H. Leptin, *Classification of C^*-algebras with a Finite Dual*, J. Functional Analysis **16** (1973) 241-257.
9. J. Bellissard, K-*theory of C^*-algebras in Solid State Physics*, in: *Statistical Mechanics and Field Theory: Mathematical Aspects*, LNP 257, T.C. Dorlas et al. eds. (Springer, 1986);
   J. Bellissard, *Ordinary Quantum-Hall Effect and Noncommutative Geometry*, in: *Localization in Disordered Systems*, Bad Schandan 1988, P. Ziesche et al. eds. (Teubner-Verlag, 1988);
   J. Bellissard, A. van Elst, H. Schulz-Baldes, *The Noncommutative Geometry of the Quantum Hall Effect*, J. Math. Phys. **30** (1994) 5373-5451.
10. N. Berline, E. Getzler, M. Vergne, *Heat Kernels and Dirac Operators* (Springer-Verlag, 1991).
11. G. Bimonte, E. Ercolessi, G. Landi, F. Lizzi, G. Sparano, P. Teotonio-Sobrinho, *Lattices and their Continuum Limits*, J. Geom. Phys. **20** (1996) 318-328; (hep-th/9507147).
   G. Bimonte, E. Ercolessi, G. Landi, F. Lizzi, G. Sparano, P. Teotonio-Sobrinho, *Noncommutative Lattices and their Continuum Limits*, J. Geom. Phys. **20** (1996) 329-348; (hep-th/9507148).

190   References

12. B. Blackadar, *K-Theory for Operator Algebras*, MSRI Publications, 5 (Springer, 1986).
13. D. Bleecker, *Gauge Theory and Variational Principles* (Addison-Wesley, 1981).
14. L. Bombelli, J. Lee, D. Meyer, R.D. Sorkin, *Space-Time as a Causal Set*, Phys. Rev. Lett. **59** (1987) 521-524.
15. J.-P Bourguignon, P. Gauduchon, *Spineurs, Opérateurs de Dirac et Variations de Métrique*, Commun. Math. Phys. **144** (1992) 581-599.
16. O. Bratteli, *Inductive Limits of Finite Dimensional C*-algebras*, Trans. Amer. Math. Soc. **171** (1972) 195-234.
17. O. Bratteli, *Structure Spaces of Approximately Finite-Dimensional C*-algebras*, J. Functional Analysis **16** (1974) 192-204.
18. O. Bratteli, G.A. Elliot, *Structure Spaces of Approximately Finite-Dimensional C*-algebras, II*, J. Functional Analysis **30** (1978) 74-82.
19. O. Bratteli, D.W. Robinson, *Operator Algebras and Quantum Statistical Mechanics*, Vol. I (Springer-Verlag, 1979).
20. N. Bourbaki, *Eléments de mathématique: algèbre*, (Diffusion C.C.L.S., 1970).
21. N. Bourbaki, *Eléments de mathématique: topologie générale*, (Diffusion C.C.L.S., 1976).
22. L.G. Brown, P. Green, M.A. Rieffel, *Stable Isomorphism and Strong Morita Equivalence of C*-algebras*, Pac. J. Math. **71** (1977) 349-363.
23. P. Budinich, A. Trautman, *The Spinorial Chessboard* (Springer-Verlag, 1988).
24. H. Cartan, S. Eilenberg, *Homological Algebra*, (Princeton University Press, 1973).
25. A.H. Chamseddine, A. Connes, *Universal Formula for Noncommutative Geometry Actions: Unification of Gravity and the Standard Model*, Phys. Rev. Lett. **24** (1996) 4868-4871;
    A.H. Chamseddine, A. Connes, *The Spectral Action Principle*, hep-th/9606001.
26. A.H. Chamseddine, G. Felder, J. Fröhlich, *Gravity in Non-Commutative Geometry*, Commun. Math. Phys. **155** (1993) 205-217.
27. A.H. Chamseddine, J. Fröhlich, O. Grandjean, *The gravitational sector in the Connes-Lott formulation of the standard model*, J. Math. Phys. **36** (1995) 6255-6275.
28. F. Cipriani, D. Guido, S. Scarlatti, *A Remark on Trace Properties of K-cycles*, J. Oper. Theory **35** (1996) 179-189.
29. A. Connes, *C*-algèbres et géométrie différentielle*, C.R. Acad. Sci. Paris Sér. A **290** (1980) 599-604.
30. A. Connes, *Non-commutative Differential Geometry*, Publ. I.H.E.S. **62** (1986) 257-360.
31. A. Connes, *The Action Functional in Non-commutative Geometry*, Commun. Math. Phys. **117** (1988) 673-683.
32. A. Connes, *Noncommutative Geometry* (Academic Press, 1994).
33. A. Connes, *Non-commutative Geometry and Physics*, in *Gravitation and Quantizations*, Les Houches, Session LVII, (Elsevier Science B.V., 1995).
34. A. Connes, *Noncommutative Geometry and Reality*, J. Math. Phys **36** (1995) 6194-6231.
35. A. Connes, *Geometry from the Spectral Point of View*, Lett. Math. Phys **34** (1995) 203-238.
36. A. Connes, *Gravity coupled with matter and the foundation of non commutative geometry*, Commun. Math. Phys. **182** (1996) 155-176.
37. A. Connes, *Brisure de symétrie spontanée et géométrie du point de vue spectral*, Séminaire Bourbaki, 48ème année, 1995-96, n. 816, Juin 1996.

38. A. Connes, J. Lott, *Particle models and noncommutative geometry*, Nucl. Phys. B (Proc. Suppl.) **B18** (1990) 29-47.
    A. Connes, J. Lott, *The Metric Aspect on Noncommutative Geometry*, in: Proceedings of the 1991 Cargèse summer school, J. Frölich et al. eds. (Plenum, 1992).
39. A. Connes, M. Rieffel, *Yang-Mills for Non-commutative Two-Tori*, in *Operator Algebras and Mathematical Physics*, Contemp. Math. **62** (1987) 237-266.
40. J. Cuntz, D. Quillen, *Algebra extension and nonsingularity*, J. Amer. Math. Soc. **8** (1995) 251-289.
41. L. Dabrowski, P.M. Hajac, G. Landi, P. Siniscalco, *Metrics and pairs of Left and Right Connections on Bimodules*, J. Math. Phys. **37** (1996) 4635-4646.
42. E.E. Demidov, *Ierarkhija Kadomceva-Petviashvili i problema Shottki* (Izdatel'stvo MK NMU 1995).
43. J. Dixmier, *Les C*-algèbres et leurs représentations* (Gauthier-Villars, 1964).
44. J. Dixmier, *Existence de traces non normals*, C.R. Acad. Sci. Paris, Ser. A-B, **262** (1966) A1107-A1108.
45. S. Doplicher, K. Fredenhagen, J.E. Roberts, *Spacetime Quantization Induced by Classical Gravity*, Phys. Lett. **B331** (1994) 39-44.
    S. Doplicher, K. Fredenhagen, J.E. Roberts, *The Quantum Structure of Spacetime at the Planck Scale and quantum Fields*, Commun. Math. Phys. **172** (1995) 187-220.
    S. Doplicher, *Quantum Spacetime*, Ann. Inst. H. Poincaré, **64** (1996) 543-553.
46. V.G. Drinfel'd, *Quantum Groups*, Proceed. Intern. Congr. Math. Berkeley, CA, 1986, (Amer. Math. Soc., 1987) pp. 798-820.
47. M. Dubois-Violette, *Dérivations et calcul différentiel non commutatif*, C. R. Acad. Sci. Paris, **I 307** (1988) 403-408.
48. M. Dubois-Violette, *Noncommutative Differential Geometry, Quantum Mechanics and Gauge theory*, in Differential Geometric Methods in Theoretical Physics, LNP 375, C. Bartocci et al. eds (Springer, 1991) 13-24.
49. M. Dubois-Violette, *Some Aspects of Noncommutative Differential Geometry*, q-alg/9511027.
50. M. Dubois-Violette, R. Kerner, J. Madore, *Gauge Bosons in a Noncommutative Geometry*, Phys. Lett. **B217** (1989) 485-488.
    M. Dubois-Violette, R. Kerner, J. Madore, *Classical Bosons in a Noncommutative Geometry*, Class. Quant. Grav. **6** (1989) 1709-1724.
    M. Dubois-Violette, R. Kerner, J. Madore, *Noncommutative Differential Geometry of Matrix Algebras*, J. Math. Phys. **31** (1990) 316-322.
    M. Dubois-Violette, R. Kerner, J. Madore, *Noncommutative Differential Geometry and New Models of gauge Theory*, J. Math. Phys. **31** (1990) 323-330.
51. M. Dubois-Violette, R. Kerner, J. Madore, *Shadow of Noncommutativity*, q-alg/9702030.
52. M. Dubois-Violette, T. Masson, *On the First Order Operators in Bimodules*, Lett. Math. Phys. **37** (1996) 467-474.
53. M. Dubois-Violette, P.W. Michor, *Dérivations et calcul différentiel non commutatif II*, C. R. Acad. Sci. Paris, **I 319** (1994) 927-931.
    M. Dubois-Violette, P.W. Michor, *The Frölicher-Nijenhuis bracket for derivation based non commutative differential forms*, ESI Preprint (1994) 133, Vienna.
54. M. Dubois-Violette, P.W. Michor, *Connections on central bimodules in noncommutative differential geometry*, J. Geom. Phys. **20** (1996) 218-232.
55. M.J. Dupré, R.M. Gillette, *Banach Bundles, Banach Modules and Automorphisms of C*-algebras*, Research Notes in Mathematics **92** (Pitman, 1983).

56. E.G. Effros, *Dimensions and C*-algebras*, CBMS Reg. Conf. Ser. in Math, no. 46, (Amer. Math. Soc., 1981).

57. E. Elizalde, *Complete Determination of the Singularity Structure of Zeta Functions*, J. Phys. **A30** (1997) 2735-2744.
    E. Elizalde, L. Vanzo, S. Zerbini, *Zeta-function Regularization, the Multiplicative Anomaly and the Wodzicki Residue*, hep-th/9701060.

58. G.A. Elliot, *On the Classification of inductive Limits of Sequences of Semisimple Finite Dimensional Algebras*, J. Algebra **38** (1976) 29-44.

59. R. Engelking, *General Topology* (Heldermann, 1989).

60. E. Ercolessi, G. Landi, P. Teotonio-Sobrinho, *K-theory of Noncommutative Lattices*, ESI Preprint (1995) 295, Vienna; q-alg/9607017.

61. E. Ercolessi, G. Landi, P. Teotonio-Sobrinho, *Noncommutative Lattices and the Algebras of their Continuous Functions*, Rev. Math. Phys., to appear; (q-alg/9607016).

62. E. Ercolessi, G. Landi, P. Teotonio-Sobrinho, *Trivial Bundles over Noncommutative Lattices*, in preparation.

63. R. Estrada, J.M. Gracia-Bondía, J.C. Várilly, *On Summability of Distributions and Spectral Geometry*, Commun. Math. Phys., to appear; (funct-an/9702001).

64. B.V.Fedosov, F. Golse, E. Schrohe, *The Noncommutative Residue for Manifolds with Boundary*, J. Func. Anal. **142** (1996) 1-31.

65. J.M.G. Fell, R.S. Doran, *Representations of *-Algebras, Locally Compact Groups and Banach *-Algebraic Bundles* (Academic Press, 1988).

66. J. Fröhlich, K. Gawedzki, *Conformal Field Theory and the Geometry of Strings*, CRM Proceedings and Lecture Notes **7** (1994) 57-97; hep-th/9310187.

67. J. Fröhlich, O. Grandjean, A. Recknagel, *Supersymmetric Quantum Theory and (Non-commutative) Differential Geometry*, Commun. Math. Phys., to appear; (hep-th/9612205).

68. J. Fröhlich, O. Grandjean, A. Recknagel, *Supersymmetric Quantum Theory, Non-commutative Geometry and Gravitation*, hep-th/9706132.

69. J. Fröhlich, T. Kerler, *Quantum Groups, Quantum Category and Quantum Field Theory*, L.N.M. 1542 (Springer-Verlag, 1993).

70. C. Gordon, D.L. Webb, S. Wolpert, *One Cannot Hear the Shape of a Drum*, Bull. Amer. Math. Soc. **27** (1992) 134-138.

71. R. Geroch, *Einstein Algebras*, Commun. Math. Phys. **26** (1972) 271-275.

72. P.B. Gilkey, *Invariance Theory, The Heat Equation, And the Atiyah-Singer Index Theorem*, 2nd edition, Studies in Advanced Mathematics (CRC Press, 1995).

73. E. Hawkins, *Hamiltonian Gravity and Noncommutative Geometry*, gr-qc/9605068.

74. J.G. Hocking, G.S. Young, *Topology* (Dover, 1961).

75. G.T. Horowitz, *Introduction to String Field Theory*, in *Superstring '87*, L. Alvarez-Gaume et al. eds. (World Scientific, 1987).

76. B. Iochum, D. Kastler, T. Schücker, *On the Universal Chamseddine-Connes Action I. Details of the Action Computation*, hep-th/9607158.
    L. Carminati, B. Iochum, D. Kastler, T. Schücker, *On Connes' New Principle of General Relativity. Can Spinors Hear the Forces of Spacetime?*, hep-th/9612228.

77. B. Iochum, T. Schücker, *Yang-Mills-Higgs Versus Connes-Lott*, Commun. Math. Phys. **178** (1996) 1-26.

78. A. Joseph, *Quantum Groups, and Their Primitive Ideals* (Springer, 1995).

79. M. Kac, *Can One Hear the Shape of a Drum?*, Amer. Math. Monthly **73** (1966) 1-23.

80. W. Kalau, M. Walze, *Gravity, Non-commutative Geometry and the Wodzicki Residue*, J. Geom. Phys. **16** (1995) 327-344.
81. W. Kalau, *Hamilton Formalism in Non-commutative Geometry*, J. Geom. Phys. **18** (1996) 349-380.
82. M. Karoubi, *K-theory: An introduction* (Springer-Verlag, 1978).
83. M. Karoubi, *Connexion, courbures et classes caracteristique en K-theorie algebrique*, Can. Math. Soc. Conf. Proc. **2** (1982) 19-27.
84. C. Kassel, *Quantum Groups*, GTM (Springer-Verlag, 1995)
85. C. Kassel, *Le résidu non commutatif [d'après M. Wodzicki*, Astérisque **177-178** (1989) 199-229.
86. D. Kastler, *A Detailed Account of Alain Connes' Version of the Standard Model*, parts I and II, Rev. Math. Phys. **5** (1993) 477-523;
    part III, Rev. Math. Phys. **8** (1996) 103-165;
    D. Kastler, T. Schücker, *A Detailed Account of Alain Connes' Version of the Standard Model*, part IV, Rev. Math. Phys. **8** (1996) 205-228.
    D. Kastler, T. Schücker, *The Standard Model à la Connes-Lott*, hep-th/9412185.
87. D. Kastler, *The Dirac Operator and Gravitation*, Commun. Math. Phys. **166** (1995) 633-643.
88. A. Kempf, *String/Quantum Gravity Motivated Uncertainty Relations and Regularisation in Field Theory*, hep-th/9612082.
    A. Kempf, G. Mangano, *Minimal Length Uncertainty Relation and Ultraviolet Regularisation*, Phys. Rev. **D55** (1997) 7909-7920.
89. J.L. Koszul, *Fiber Bundles and Differential Geometry* TIFR Publications (Bombay, 1960).
90. T. Kopf, *Spectral Geometry and Causality*, gr-qc/9609050.
91. G. Landi, *An Algebraic Setting for Gauge Theories*, PhD Thesis, SISSA Trieste, December 1988, unpublished.
92. G. Landi, G. Marmo, *Algebraic Differential Calculus for Gauge Theories*, Nucl. Phys. (Proc. Suppl.) **18A** (1990) 171-206.
93. G. Landi, C. Rovelli, *General Relativity in Terms of Dirac Eigenvalues*, Phys. Rev. Lett. **78** (1997) 3051-54; (gr-qc/9612034).
94. G. Landi, C. Rovelli, *Kinematics and Dynamics of General Relativity with Noncommutative Geometry*, in preparation.
95. G. Landi, Nguyen A.V., K.C. Wali, *Gravity and Electromagnetism in Noncommutative Geometry*, Phys. Lett. **B326** (1994) 45-50.
96. H.B. Lawson, M.-L. Michelsohn, *Spin Geometry* (Princeton University Press, 1989).
97. F. Lizzi, G. Mangano, G. Miele, G. Sparano, *Constraints on Unified Gauge Theories from Noncommutative Geometry*, Mod. Phys. Lett. **A11** (1996) 2561-2572.
98. F. Lizzi, G. Mangano, G. Miele, G. Sparano, *Fermion Hilbert Space and Fermion Doubling in the Noncommutative Geometry Approach to Gauge Theories*, Phys. Rev. **D55** (1997) 6357-6366.
    F. Lizzi, G. Mangano, G. Miele, G. Sparano, *Mirror Fermions in Noncommutative Geometry*, hep-th/9704184.
99. S. Majid, *Foundations of Quantum Group Theory* (Cambridge University Press, 1995).
100. J. Madore, *An Introduction to Noncommutative Differential Geometry and its Physical Applications*, LMS Lecture Notes 206, 1995.
101. Y.I. Manin, *Quantum Groups and Noncommutative Geometry*, Centre Rech. Math. Univ. Montréal, Montréal, Quebec, 1988.

102. C.P. Martín, J.M. Gracia-Bondía, J.C. Várilly, *The Standard Model as a non-commutative geometry: the low mass regime*, Phys. Rep., to appear; (hep-th/9605001).

103. T. Masson, *Géométrie non commutative et applications à la théorie des champs*, PhD Thesis; ESI Preprint (1996) 296, Vienna.

104. J. Mickelsson, *Wodzicki Residue and Anomalies of Current Algebras*, hep-th/9404093.

105. J. Milnor, *Eigenvalues of the Laplace Operator on Certain Manifolds*, Proc. Nat. Acad. Sci. USA **51** (1964) 775.

106. A.S. Mishchenko, *C^*-algebras and K-theory*, in *Algebraic Topology*, Aarhus 1978, J.L. Dupont et al. eds., LNM 763 (Springer-Verlag, 1979).

107. J. Mourad, *Linear Connections in Noncommutative Geometry*, Class. Quant. Grav. **12** (1995) 965-974.

108. G. Murphy, *C^*-algebras and Operator Theory* (Academic Press, 1990).

109. W. Paschke, Inner Product Modules over B^*-algebras, Trans. Amer. Math. Soc. **182** (1973) 443-468.

110. G.K. Pedersen, *C^*-algebras and their Automorphism Groups* (Academic Press, 1979).

111. L. Pittner, *Algebraic Foundations of Non-commutative Differential Geometry and Quantum Groups*, LNM m39 (Springer, 1995).

112. R.J. Plymen, *Strong Morita equivalence, Spinors and Symplectic Spinors*, J. Oper. Theory **16** (1986) 305-324.

113. M.A. Rieffel, *Induced Representation of C^*-algebras*, Bull. Amer. Math. Soc. **78** (1972) 606-609.
M.A. Rieffel, *Induced Representation of C^*-algebras*, Adv. Math. **13** (1974) 176-257.

114. M.A. Rieffel, *Morita Equivalence for Operator Algebras*, in *Operator Algebras and Applications*, Proc. Symp. Pure Math. **38**, R.V. Kadison ed. (American Mathematical Society, 1982) 285-298.

115. M.A. Rieffel, *C^*-algebras Associated with Irrational Rotations*, Pacific J. Math. **93** (1981) 415-429.
M.A. Rieffel, *Non-commutative Tori: a Case Study of Non-commutative Differential Manifolds*, in *Geometric and Topological Invariants of Elliptic Operators*, Contemp. Math. **105** (1990) 191-211.

116. M.A. Rieffel, *Critical Points of Yang-Mills for Noncommutative Two-Tori*, J. Diff. Geom. **31** (1990) 535-546.

117. M. Reed, B. Simon, *II : Fourier Analysis, Self-Adjointness* (Academic Press, 1975).

118. W. Rudin *Real and Complex Analysis* (McGraw-Hill, 1987).

119. A.D. Sakharov, *Vacuum Quantum Fluctuations in Curved Space and the Theory of Gravitation*, Sov. Phys. Dokl. **12** (1968) 1040-1041.
A.D. Sakharov, *Spectral Density of Eigenvalues of the Wave Equation and Vacuum Polarization*, Theor. Math. Phys. (USSR) **23** (1976) 435-444.

120. B. Simon, *Trace Ideals and their Applications*, London Mathematical Society Lecture Notes 35 (Cambridge University Press, 1979).

121. R.D. Sorkin, *Finitary Substitute for Continuous Topology*, Int. J. Theor. Phys. **30** (1991) 923.

122. R.P. Stanley, *Enumerative Combinatorics*, Vol. I (Wadsworth & Brooks/Cole, 1986).

123. R.G. Swan, *Vector Bundles and Projective Modules*, Trans. Am. Math. Soc. **105** (1962) 264-277.

124. M.E. Sweedler, *Hopf Algebras* (W.A. Benjamin, 1969).

125. T. Schücker, J.-M.Zylinski, *Connes' Model building Kit*, J. Geom. Phys. **16** (1995) 207-236.

126. M.E. Taylor, *Pseudodifferential Operators* (Princeton University Press, 1981).

127. W. Thirring, *Classical Field Theory* (Springer-Verlag, 1985).

128. A. Trautman, *On Gauge Transformations and Symmetries*, Bull. Acad. Polon. Sci., Ser. sci. phys. et astron. **27** (1979) 7-13.

129. H. Upmeier, *Jordan Algebras in Analysis, Operator Theory and Quantum Mechanics*, CBMS Reg. Conf. Ser. in Math, no. 67, (Amer. Math. Soc., 1987).

130. J.C. Várilly, J.M. Gracia-Bondía, *Connes' Noncommutative Differential Geometry and the Standard Model*, J. Geom. Phys. **12** (1993) 223-301.

131. N.E. Wegge-Olsen, *K-theory and C^*-algebras* (Oxford Science Publications, 1993).

132. J. Wess, B. Zumino, *Covariant Differential Calculus on the Quantum Hyperplane*, Nucl. Phys. B (Proc. Supp.) **18B** (1990) 302-312.

133. E. Witten, *Non-Commutative Geometry and String Field Theory*, Nucl. Phys. **B268** (1986) 253-294.

134. M. Wodzicki, *Local Invariants of Spectral Asymmetry*, Inv. Math. **75** (1984) 143-177.

     M. Wodzicki, *Noncommutative Residue*, In *K-theory, Arithmetic and Geometry*, Yu. I. Manin ed., LNM 1289 (Springer, 1987).

135. S.L. Woronowicz, *Compact Matrix Pseudogroups*, **111** (1987) 631-665.

136. R. Wulkenhaar, *A Tour Through Nonassociative Geometry*, hep-th/9607086.

     R. Wulkenhaar, *The Mathematical Footing of Nonassociative Geometry*, hep-th/9607094.

     R. Wulkenhaar, *Grand Unification in Nonassociative Geometry*, hep-th/9607237.

     R. Wulkenhaar, *The Standard Model Within Nonassociative Geometry*, Phys. Lett. **B390** (1997) 119-127.

Index